THE PAY PER CALL REVOLUTION

How an Elite Group of Performance
Marketers Are Taking Control
and Building Highly Profitable
Businesses on Their Terms

ADAM YOUNG

Copyright © 2024 by Adam Young

All rights reserved. No part of this publication may be reproduced, distributed, or transmitted in any form or by any means, including photocopying, recording, or other electronic or mechanical methods, without the prior written permission of the publisher, except in the case of brief questions embodied in critical reviews and certain other noncommercial uses permitted by copy law. For permission requests, write to the publisher at the website address below.

Performance Marketing Media

Paperback ISBN: 979-8-9892471-0-3
Hardcover ISBN: 979-8-9892471-2-7
Ebook ISBN: 979-8-9892471-1-0

Interior Layout by: Amit Dey
Index by: Valerie Haynes Perry

This book is intended to provide information about utilizing software tools for potential financial gain and is for educational purposes only. It is important to understand that every individual's results will inherently differ based on various factors specific to their circumstances, such as their business model, strategies employed, and the nature of their products or services.

We do not guarantee or warrant any specific level of success, income, or sales. The software discussed in this book is designed to assist in creating opportunities, yet its effectiveness may vary depending on external market factors beyond our control and it may not be applicable to every business.

This book does not present a business opportunity, a "get rich quick" scheme, business in a box, or a guaranteed system for financial gains. We do not make claims regarding specific earnings, efforts, or return on investment. Potential outcomes and profitability are subject to your efforts, dedication, and the uniqueness of your business. In fact, you may lose money by attempting anything talked about in this book.

Please be advised that the information provided herein is not intended to serve as tax, accounting, financial, or legal advice. Consultation with qualified professionals such as your accountant, attorney, or financial advisor is recommended for personalized guidance in these areas.

By using the guidance and information in this book, you accept that results may vary, you will likely lose money if you attempt anything discussed in this book, and success is dependent on individual efforts and market conditions. Ringba, LLC and Adam Young disclaim any liability arising from the use of this information for any purposes.

CONTENTS

Dear Reader v
Introduction xi

1. What Is Pay Per Call Anyway? 1
2. How Pay Per Call Works 13
3. The Simplest Way to Get Started 35
4. Choosing Verticals 49
5. Building a Professional Brand 63
6. Building Your First Campaign Step by Step 73
7. Compliance and Quality Assurance 83
8. The Psychology of Advertising 95
9. Digital Traffic Sources 113
10. Print Advertising 131
11. Out-Of-Home Advertising 149
12. Radio Advertising 161
13. Television Advertising 175
14. Building a Buyer Network 189
15. Negotiating for Scale 205
16. Ringba Secrets: Optimizing Call Flows with
    Automated Call Routing 217

17. Why Affiliates Fail 241
18. The Fastest Way to Win. 247

Conclusion 253
Pay Per Caller Manifesto 257
About Ringba 259
Endnotes. 261
About the Author 263
Resources 265
Glossary. 267
Index. 279
Testimonials 287

DEAR READER

You are incredible. You have the heart and spirit of an entrepreneur. You don't let challenges get in your way. You're not afraid to build and fail and build again. And that is something unique and special in this crazy world of ours.

You have the potential to change the world!

Has anyone ever told you that? And honestly meant it?

If not, trust me, you're not alone.

If you're anything like I was, maybe you've faced a lifetime of rejection and ridicule, bullying and failure.

Maybe you're going through that right now and feel like there's no one out there who will ever believe in you.

Maybe you've turned to risky behaviors like drugs and alcohol to simply feel better for a little while.

Maybe you've hit rock bottom, only to find there were even lower levels of suck you didn't know about.

And maybe you've also felt the highest of highs when you crushed a campaign or created a seven-figure launch with nothing but the thoughts in your head and a beat-up laptop.

Those are the good times, right? The times when you can tell your haters, your teachers, your so-called friends, and anyone else who wrote you off as a loser—you can just tell them all to kiss it! You are not a loser. You weren't wasting time all those years you spent messing around online.

You were learning. You were absorbing. You were trying and failing and trying again.

That's called grit.

You've picked yourself up more times than you can count, and every time it gets better. You get smarter. You've learned to navigate the rough waters of performance marketing.

I've been there. I've made and lost more money than most people could ever dream of. I've sold door-to-door in the frigid Michigan winter with a massive hole in my shoe and had to wring out my ice-cold soaking wet socks between appointments.

I built the world's shittiest business—and lived to tell the tale.

I built amazingly successful campaigns that disappeared overnight.

I've lived in my parent's basement not once, not twice, but ... well, I'll let you guess how many times I moved back into that windowless hole in the ground.

Maybe you can relate.

In the pre-dawn era, before people even knew what affiliates were, I tried *everything* to make money online.

I started my own college joke site.

I tried trading traffic and promoting affiliate products on blogs.

I tried drop shipping, selling phone equipment, and selling distributor products on websites.

I tried building SEO sites for ad clicks.

Everything failed.

And when a business did finally succeed, something completely out of my control would bring the whole thing crashing down again.

I call it "the affiliate curse."

If you've ever worked in affiliate or performance marketing, you know what I mean. You can be the best media buyer in the history of media buyers, and all it takes is one advertiser to decide they're done and your whole livelihood goes up in smoke. No warning. No apologies. You're just done. And if you're married or have kids, heaven help you. The stress is unbelievable.

But your skills are still there. And there's always another offer that promises to be "the one." Am I right? This offer won't abuse you like the last one did. This time you'll build a business that will last. This time will be different.

I'm here to tell you, I get it. I've been there.

And I want to show you a different world. I want to introduce you to a brand-new paradigm in performance marketing that will change your life.

It's not some crazy new offer that's sure to take off and "never die."

It's not a ground-floor opportunity you'll "never regret" investing in.

In fact, it's the opposite.

What if I told you that you could take those marketing skills you've acquired with blood, sweat, and tears, and put them to work for companies that are household names? I'm talking about giant categories like auto insurance, pest control, satellite and cable TV, Medicare and hundreds of others. Would you like to be on their marketing teams?

These industries aren't going *anywhere,* and they need your expertise. And the best part is they don't want to hire you. They just want to pay you to drive traffic to their call centers.

My goal with this book is to open your eyes to a new way of building a business as a marketer—a business that's created on a solid foundation of real-world companies that will be here for decades to come. I want you to see how you can utilize your unique skills to actually help people, impact the world, and make incredible amounts of money at the same time.

So, whether you're currently riding high on a series of amazing offers but aren't really secure because you know your business is built like a house of cards ...

Or you're sitting at rock bottom after those cards collapsed, again, and you're looking at moving back into your parents' basement, again...

I want you to know there's hope.

You're incredible.

I believe in you.

The most amazing future is within your grasp. You are so close—this book will show you the way. All you have to do is read it.

INTRODUCTION

Recently, I was sitting in the dark with about 15,000 other people staring at a brightly lit stage. At one point, a hush came over the crowd and the artist on stage told everyone to pull out their phones and turn on the flashlight. In a few seconds, the entire stadium was lit up with what seemed like tens of thousands of fireflies dancing in the air. At that moment it hit me, everyone has a phone.

Of course I knew this before I stood there watching all those lights twinkle, but I didn't really comprehend what it meant. Every single human has one of these things in their pocket almost every waking hour of the day, and all we have to do is get them excited enough to click "call."

The coolest thing about a phone call is that it has the highest level of consumer intent. It's not like a click or a lead where you've captured their attention for a second or maybe a couple minutes. When a consumer picks up the phone and calls they have *already considered buying* and only need their questions answered to complete a transaction. This means that phone calls are worth significantly more money to a business than a click or a lead. A phone call guarantees them a real chance at acquiring a new customer for almost any amount of money without having to do any more work.

Not that there's anything wrong with clicks or leads. I certainly have made a significant amount of money running these types of campaigns. But when businesses buy a click, they have to turn that click into a lead, and call that lead numerous times to connect with a consumer. Sometimes they get the consumer on the phone, sometimes they don't, but the consumer will never be as excited about whatever they're selling as they were the moment they clicked or submitted the form. It's also getting harder and harder for businesses to dial consumers even if they are interested. I don't know about you, but my phone rings off the hook, and I don't answer phone calls from numbers I don't recognize anymore.

At first, when I discovered pay per call, it blew my mind how much a single phone call could be worth. After time though, I realized it makes complete sense. A business is willing to pay more money for an inbound phone call than a click or a lead because the consumer is interested in buying right then.

Not only that, businesses also aren't going to bother answering the phone for a few dollars. They only invest in paying for phone calls when they're selling a high-value product or a service someone urgently needs.

Just think about someone getting a flat tire on the expressway. I don't know about you, but I'm not interested in risking my life to save a few bucks changing a dirty tire while cars are whizzing by at 70 miles an hour. Most people don't know how to change a tire, and even if they do, it isn't very fun.

So, what do they do? They pull out their phone and search for a tow truck right then and there. They don't fill out a lead form or click to read a blog. They don't want to wait for someone to get back to them. They call the towing company immediately for the fastest service possible. What many consumers don't know is that it's highly likely that someone made a nice commission just because they called that number.

Or perhaps someone needs insurance for their house or car, but they have a lot of questions. To get their questions answered, they need to speak with a licensed agent over the phone, and only a licensed agent can sell them a policy. While many people may click around to shop on the internet, in order to find their best options, they need to speak with an expert.

No one wants to fill out a form and wait for someone to get back to them, and they definitely don't want their phone ringing off the hook while people compete for their business.

They want to solve their problem or desire immediately without all the hassles. That is why most of the time they just pick up the phone and call. And if it's your number they call, you can get paid for it.

Considering the amount of money generated, the pay per call industry is still relatively small. There are really only a few thousand people that make up the entire ecosystem of buyers, networks, and affiliates.

That is why what you're about to read in this book is so special. The industry is in its infancy, so there is an astronomical amount of opportunity available for those willing to learn the secrets.

A few weeks before I started writing this book, I asked my team to pull the year's biggest campaigns from Ringba, my call-tracking software platform. My jaw dropped when I saw that one of our customers had run $90,299,124.54 on a single campaign. In all my years of affiliate marketing I had never seen anything like it, and it speaks to exactly why pay per call is a massive opportunity.

But here's the thing ...

Pay per call isn't like traditional affiliate marketing; it's a whole different ball game. And if you want to play, you can't do things the way you're used to. That's one reason I decided to write this book. You gotta know the rules. And how are you going to learn the rules if no one is teaching them?

Here's the other thing ...

The pay per call space may be small now, but it won't be that way forever. There's a revolution coming, and the entire way people buy and sell products and services will be turned on its head. Will you be on the leading edge of that wave?

I hope by the end of this book you'll say, "Hell yeah, Adam! I'm in."

CHAPTER ONE

WHAT IS PAY PER CALL ANYWAY?

When I was 10 years old, my father bought our first computer—an Intel 386. He brought it downstairs to our then-unfinished basement in our typical middle-class Michigan home, set it down on a piece of plywood on the freezing concrete floor and said, "I just bought this thing. It's going to be your future, so you need to figure it out." Then he tramped back upstairs without another word.

I was fascinated by this thing. I pulled it out of the box, figured out how to put it all together, and when I turned it on, it clicked and whirred. It had fans and janky software. It ran on Windows 3.1 and had almost nothing preinstalled. Even the CD-ROM drive was finicky. I had to quickly open and close it to get the disc to load and kept a bent paperclip handy for those times the button didn't work. At the time, there were no computer books and certainly no internet as we know it today, so I did what any smart, curious 10-year-old would do—I found the file explorer and started running every single file.

It was like Disneyland in pixels and programming. I didn't know how to read assembly language. I didn't understand what configuration files were. And I broke that computer so many times, it was unbelievable. All I wanted to do was spend every waking hour in the basement playing with that computer.

It's safe to say I was obsessed.

Then one day I convinced my parents to buy a modem, and that changed everything. Once I plugged it into the phone line, they were never able to use the phone again. I was on AOL and bulletin boards 24 hours a day. My favorite hangout spots were the programming chat rooms where people were making software in C and Visual Basic. Nobody knew or cared that I was just a kid. Profiles didn't exist back then. We just used our screen names, so there was complete anonymity. I got to learn all the software tricks and hacks from actual professionals.

That was an incredible learning experience for me. I could literally ask total strangers for help, and they would help me. No questions asked. That taught me that if you're friendly and fun, people will generally tell you anything you want to know.

One day, I decided I wanted to figure out how to make money on the World Wide Web. So I took my life savings (which wasn't much for a 10 year old) and paid monthly for a dedicated server. Obviously, I didn't have an actual business, so I just signed up with our home address and phone number. My plan was to go into the AOL chat rooms and get people to pay me $10 a month for hosting a website. Of course, nobody wanted to pay because they could get hosting for free on Geocities. But one day I got a bite. They wanted my $25 package. Sweet!

So like a week later, this check shows up, I deposit it, and I'm in business. They're paying me $25 a month, and I know I'm going to be rich. All I need is like 10 more of those people, and I'd have $250 a month. That's like millions to a child in the early '90s. My parents were flabbergasted that I'd convinced a stranger to send me money, but I hadn't done anything wrong, so they just let me keep going. Everything was set up, and my first customer was online.

Then, at 4 a.m. the next morning, our home phone starts ringing off the hook. Literally, it wouldn't stop ringing. My parents barged into my room, screaming and yelling. People were calling our house and accusing them of committing a crime and saying we were going to jail.

Apparently, I had sold my first hosting package to an email spammer. Huh. Go figure.

Now, in my defense, email spam wasn't even a thing at the time, and it certainly wasn't a crime. I didn't even know what it was. But that didn't stop the phone from ringing for hours and my server from being shut down. My parents were *pissed*! To this day I'm not sure how I managed to convince them to let me keep the computer and the internet.

But that was my first experience with making money—and getting burned—online.

It would not be my last.

I moved out of that cramped, dark, freezing basement three times in my adult life.

Every time I waved goodbye, I thought *this is it. I've figured it out. Nothing can stop me now.*

But the internet is a cruel mistress, and it felt like I had to learn everything the hard way.

Maybe you can relate.

Maybe you're one of the millions of people in search of money and freedom online. You know those people who will gladly work like crazy today for the promise of a better life in the future. And when they finally get close or even if they reach the goal, the rug is pulled out from under them. Offers close down. Algorithms change. Suppliers dry up. White hat becomes black hat overnight.

They are *never* secure—they're always one "slap" away.

Here's a fun game. What would performance marketing look like if affiliates ran the industry and marketers had complete control over their destinies?

It would mean they could pick and choose from established partners that aren't going anywhere instead of random offers by startups and less reliable companies.

It would mean being able to use dozens of different marketing mediums, not just a handful of digital ones that everyone else is also using.

It would mean complete transparency in every campaign, including recordings of every conversion.

It would mean having lots of advertisers to choose from and being able to switch traffic between them at the touch of a button.

It would mean exponentially more money, for less work in the long run.

It would mean campaigns designed to scale and last for years, not for a few months.

It would mean the potential for true set-and-forget campaigns instead of monitoring performance 24/7 and constantly adapting for every blip on a screen.

It would mean no one could rip you off, so your work is truly yours.

It would mean building enterprise value so the business is worth real money when the time comes to retire.

All that sounds like a dream, doesn't it? But that's not reality. Marketers have been forced to just accept that the business runs a certain way and there's nothing they can do about it.

Being a publisher in the performance marketing space is like sitting on the floor and building an elaborate castle with some other kid's blocks. At any time, they can decide to take their blocks back and go home. Anyone can come along and kick your beautiful creation to the ground. The more elaborate your castle, the bigger the mess when it all comes crashing down.

Here's a crazy thought …

What if you could build a business with your own blocks?

There's a big difference between being a marketer and being a business owner who learns marketing. If you're a business owner, you have to master a dozen different skills and keep them balanced to move the organization forward. People praise you for your determination and grit. They can see your efforts and support you all the way.

But being a marketer is weird. People look down on you. Your own family will ask you when you're going to get a real job. They just don't get it.

Marketers have the most valuable skill set on the planet! But it's worthless without something to sell. So, as a marketer, you either have to work for someone else, market their products, and make them millions of dollars, or you start some random business, perhaps even open an agency, and use your marketing skills to make it grow. The problem is there are so many other dynamics involved in running a successful business, not the least of which is having a decent product. You wind up doing everything except what you really love—marketing.

That leaves the performance space, which seems amazing at first. I mean, you get to use your skills and do the fun stuff while someone else deals with product development, delivery, and customer service. What's not to love?

You sign up for an offer and start driving traffic. Maybe you choose organic traffic first because it's basically free, except for your time, and you're still eating ramen in your parents' basement. Sooner or later, though, you figure out that it takes *forever* to get any traction with organic methods.

So, you start building ad campaigns and finally get some momentum going. You did it! You're finally on Easy Street. Then something happens. Maybe you run out of money. Maybe the offer crashes. Maybe a global catastrophe wipes you out.

It's fine. Just start over. You know how to do this. It'll be fine.

Over and over, round and round you go—never quite where you want to be.

That's where I was, always thinking…

I just need to spend more.

I just need to learn this one new platform.

I just need a better offer to promote.

The affiliate model seems sexy and awesome when you first come across it. Who wouldn't want to get paid to promote something you don't have to create, deliver, or manage the customer service for? You get to focus on driving traffic, which is all you really want to do anyway. All you have to do is pick an offer, and you're off to the races. Sweet!

But here's the thing we all find out sooner or later: The traditional affiliate marketing world is kind of abusive. The whole model is built in favor of the networks and the advertisers. They are at the top with an army of affiliates to support them. There's no incentive for them to care about you because there are always plenty more marketers to take your place.

You are expendable, and therefore you have to follow the rules. And when you get burned, you have little to no recourse. All you can do is start over.

That's just the way things are in this industry.

Or so you thought ...

What if I told you there's a world where affiliates have the power and infrastructure to go out and get their own clients? And they could even consider becoming their own network if they choose.

What if I told you there's a world where affiliates get paid just for driving qualified traffic *regardless* of whether a sale is made or not?

What if I told you these affiliates promote offers that people need right away, so there's real built-in urgency. They don't have to convince anyone of anything. Just show up at the right time with a solution.

And what if no one ever ripped them off, copied their ads, or stole their traffic?

Welcome to pay per call

The pay per call model turns the unfair and abusive affiliate model on its head. Instead of the advertisers having all the control, you do. Pay per call marketers build campaigns and drive traffic for an entire industry, not just individual offers. They may be in the insurance space or home services or financial—there are hundreds of verticals to choose from. But instead of only having a single advertiser to send that traffic to, there are, perhaps, dozens. If one misbehaves or goes away, the pay per caller pushes a button and sends that traffic to another advertiser. No sweat. No crashes. No more bullies.

What Is Pay Per Call Anyway?   7

Traditional affiliate: Marketers get paid if there's a sale, so they have to be great at driving traffic, pre-framing, and helping to sell the offer.

Pay per call: Marketers get paid for each qualified call *regardless* of whether a purchase is made, so they can focus on the one thing they do best—driving traffic.

In this marketing game you can push and push, hustle and grind, and burn yourself out for no reason other than it's what everybody else does. People in this space focus on the beginning of the journey—the hardest part—because so few people ever really figure it out and build a long-term sustainable career out of performance offers. There are precious few who have "survived the wars" and lived to talk about it, and even fewer who actually want to talk about it. Successful lifelong affiliates tend to quietly go about their business, not drawing attention to themselves.

But I have been there, and I do want to talk about it. I want to scream it at the top of my lungs to anyone who will listen: THERE IS ANOTHER WAY!

That survivor spirit is glorified in the marketing world. Like a boxer who just keeps getting knocked down, you're only a true winner if you keep getting back up and taking more abuse. The marketer's life goes up, down, up, down until they've had enough crashes to finally listen to their parents, quit this crazy ride, and go get a real job.

I drank that Kool-Aid. I fell for the quick cash infusions and the easy money— and fell *hard*. I had incredible highs, and, damn, did it hurt when everything crashed around me. One week I'm literally living in the princess of Belgium's castle with my best bros, and the next week I'm back in my parents' basement. Again.

I wasn't smart enough to say *fuck it* and go get a real job. I kept believing that I must have missed something critical and that I could fix whatever went wrong. I wouldn't make the same mistake the next time.

Does that sound familiar? Do you blame yourself for losing everything? Like you did something wrong?

Do you ever feel like if you could just pick the right offer, master another tactic, or spend more on ads that you'd finally come out ahead once and for all? Maybe that's true. It's possible that you have more to learn.

But maybe it's not you. Maybe the industry is abusing you, and you secretly love it because it gets you what you need. It lets you be the underdog, the scrappy fighter who never gives up. Maybe you have a touch of Stockholm Syndrome.

I know a thing or two about abuse. I was bullied every freakin' day throughout my childhood. Kids bullied me. Their parents bullied me. Coaches and teachers had no idea what to do with me. I was the nerdy kid from the swim team with a violin and a space between my front teeth that you could drive a truck through who everyone called "Gaps." I was in survival mode for 90 percent of my childhood. (OK, maybe a bit more than my childhood.)

Now, take a deep breath ...

And let me show you something different.

The pay per call world is an alternate universe.

You are in complete control of your own destiny.

There are no crashes (unless they're self-inflicted).

It might take a little more time to build your beautiful castle, but no one is going to kick it down and steal your blocks.

Plus, there's always another level to expand to. You don't reach the top and have to sabotage yourself just to find the next challenge. There are always new advertising mediums to explore and new levels of business to conquer. It's a never-ending upward spiral with continuing opportunities for personal and business development. When you're not fighting against the traditional affiliate model, you can work more creatively and actually enjoy the journey every step of the way.

The beautiful thing about pay per call is you have so many ways to advertise, and most of them have practically no competition. You'll probably start with online marketing and digital traffic sources. That's great! Online marketing has a very low barrier to entry. It's cheap, it's easy once you get the hang of it—and it's fast! You can have campaigns up and running in hours. You can test ideas in real time. And you get practically instant analytics.

But as great as all that is, there are some drawbacks.

- It's cheap to get started, but you have no control over what the ads actually cost. You can't negotiate with Zuckerberg.
- It's easy to build, but it's also easy for competitors to rip you off.

- You're always one "slap" away from an algorithm pulling the rug out from under you.
- It's a bloody red ocean full of competitors.

Most marketers these days have no idea that there's a whole world of marketing aside from the internet. They've been gaslit into thinking that online marketing is the only thing that works. If that's so, why do the biggest brands in the world still use TV, radio, direct mail, and other media to sell their products and services?

Maybe you've heard those other mediums are dead. They don't work anymore. They're only for branding. They're too expensive. There are limited targeting opportunities. Or it's just too hard to make a buck.

I call bullshit.

It's true that offline marketing mediums take time to set up. They take even longer to test. And you do need more money up front to get started—but not as much as you might think.

It's also true that the big companies use offline mediums for branding, but that doesn't mean they don't work for direct response. It just means that not very many direct response marketers know about these amazing opportunities. After all, there aren't many gurus teaching a $997 course on how to get rich in radio.

The way I see it, why not use all the tools in the toolbox?

For every 1,000 affiliates fighting it out on Facebook, there's a guy I know making eight figures in pay per call with nothing more than newspaper classified ads.

For every 1,000 affiliates trying to leverage AI to squeak out an advantage, there's a pay per caller quietly killing it on radio.

This book is filled with both modern online traffic strategies *and* proven old-school strategies that work like crazy if you're willing to do the work. You'll get the low-barrier-to-entry options and some that take a little more experience and cash. The point is you aren't locked in to one strategy that everyone else is using. You get to actually choose what you like to do.

My goal with this book is to open your eyes to a new world of opportunity. I watch performance marketers get beat up every day, and they keep going back for more because they don't know what else to do. I know. I've been there.

Full disclosure: I no longer run pay per call campaigns. Here's why: I made a ton of money in the pay per call industry, but I knew I could do so much more if I had certain tools. I needed to be able to track my calls and traffic sources. I needed to be able to listen to every call for quality assurance purposes. I needed to be able to easily reroute calls if one buyer pulled out or misbehaved. I needed to be able to target based on more than just area codes.

I knew what I needed, but there were no tools on the market that could do any of it.

So I built my own.

These days, I spend my time developing Ringba into the best damned call-tracking platform on the planet and evangelizing the message that you can walk away from an abusive industry. Pay per call is the way.

Do I hope you'll use Ringba to build your business in pay per call? Sure. Of course I do. There's no better way to keep 100 percent control of your own destiny. But is that my primary motive? Not at all.

When I was a kid, I couldn't help myself when I got beat up for being smart and different. But now that I'm older, I can help others. That's my motivation.

If you're tired of having the rug pulled out from under you...

If you're exhausted from competing in a red ocean...

If you can't stand to start over even one more time...

This book was written for you.

There's a revolution coming.

Welcome to the New World!

CHAPTER TWO

HOW PAY PER CALL WORKS

I walked into the living room with a fresh bowl of Cheerios and my heart sank. I ignored the phone buzzing in my pocket as I stared at six giant blank screens. My mind grasped at every straw it could think of. Maybe the internet was out. Maybe our advertiser was just down. Hopefully our server cluster had just crashed. (You know you've entered a dark place when you're praying for an Amazon outage.)

Normally, those screens would be scrolling with so much click activity, it's like watching *The Matrix* at 10x speed. But now they were completely still. It was eerie. My phone buzzed again. I sat down on the floor and answered.

"It's over," my business partner, Harrison, said.

Apparently, our advertiser decided to pull the plug without warning in the middle of the night. They weren't able to monetize the traffic anymore, so instead of trying to optimize the offer, they decided to call it quits. Everything we had been building for the previous two years was gone in a blink of an eye.

I think at that moment I almost had a breakdown. I didn't even notice the cereal spilling down my leg. I couldn't believe this was happening. Again.

That's the thing about affiliate marketing, the ads burn out, the offers die, the advertisers go under, this is business as usual in our world. It's the dreaded affiliate curse. That day was my lowest point. I was sick of it. I wanted out.

Over the previous 10 years, I had promoted everything under the sun across every major affiliate network in the industry and had become a master of affiliate marketing and traffic arbitrage. I learned how to run just about any type of campaign on any paid traffic source, but no matter how good things got, I knew it wouldn't last, and eventually I would have to start from scratch over and over.

For a long time I just shrugged it off, chalked it up to "that's just the business of affiliate marketing," and started over. I did that hundreds of times. But this time, I was done. I was sick and tired of having no control over my own destiny.

We made plenty of money. But we hadn't *built* anything lasting. We weren't helping people. We weren't creating opportunities. We weren't doing anything that would change the world. It was just a puzzle. I like puzzles, but there was no fulfillment in it.

That day, my pants soggy from spilled milk, I quit. Just like thousands of affiliates before me, I reached the end of my rope. Fortunately for me, I didn't have to sacrifice all the skills I'd built up over the years. My expertise still counted for something. I just had to find a different way to leverage those skills.

So Harrison and I started looking for our next opportunity. Not some fly-by-night offer this time but a real opportunity. One that helped people and would give us the security of a rock-solid business model, one that wouldn't disappear in a few months, leaving us stranded.

Later that month, I discovered pay per call.

Why pay per call is superior to all other types of offers

Phone calls carry the highest buying intent of any consumer action. Someone may search online for "pest control near me," but they could just be browsing companies, comparing prices, and checking out reviews. Gone are the days when people made calls to find out that information. It's all online. And we've been trained to do that initial research by ourselves.

People only make phone calls if they're actually interested in purchasing a particular product or service. They know that if they make that call, they have the expectation that they're going to pull out a credit card and buy right away.

That's why businesses get excited about phone calls and why they're willing to pay expert marketers to just send them calls: so they can focus on what they're great at—delivering their own products and services.

If a company buying phone calls is paying $10 per call and has an average close ratio of 25%, they know their cost of acquiring a customer is approximately $40 every single time. Which do you think is more cost effective: spending years learning advertising and paying for a full-time marketing department or ad agency, or just handing over a little money to an expert by the call?

As a marketer, you understand how to calculate the cost of acquisition for any particular customer, but most business owners don't have a clue. So pay per call allows them to understand exactly how much they're spending per lead, which gives them the confidence to keep paying for those calls.

They don't have to take the risk of new advertising campaigns, learning new promotional methods, creating new promotional materials, or doing any other marketing tasks. They can simply focus on their business and leave all of the marketing and user acquisition to their partners.

And that's where you come in. That fixed-call cost creates a huge opportunity for skilled marketers like you to come up with new ways to promote the campaign. And since the advertiser is never aware of the cost of acquisition, they have no idea what your profit margin may be.

Once a successful campaign is up and running, you reap the benefits as long as the advertiser is willing to buy the calls or as long as the campaign keeps running. Maybe generating a call costs you $2, but you're selling it to a grateful buyer for $8 to $25. That's a pretty sweet margin.

Also, consider how much is going on in call centers all over the world. They are in chaos most of the time. The managers have the insane task of trying to keep every agent on the phone closing deals as fast as possible all day long. It's a goal they rarely reach. Agents spend 25% of their time sitting idle. The turnover rate for agents is 38% as of this writing. And it costs between $5,000 and $25,000 to replace a single agent who may make only $12 an hour.[1]

Can you even imagine the stress? So by taking on the job of driving as many calls to those centers as possible, you are helping those managers immensely. You're making their jobs infinitely easier, which means the tension in the call center is lower, which means the agents have a better work environment, which means there's less attrition and everyone is happier. All because you know how to drive traffic.

It's an amazing advertising model. You're making a great profit, and the buyer knows exactly what it costs them to acquire each and every lead. It's a win-win situation. You can't beat that.

The first time I really understood the power of this model I was hanging out with Harrison and a friend in the presidential suite at the Waldorf Astoria in New York City. My friend told me he was running these amazing pay per call campaigns. I had never seen one of these campaigns before, and I didn't quite understand what he was saying, so I asked him to show me.

A few clicks of the keyboard later and we were looking at his tracking platform. He showed me how many phone calls had run through the platform and how many minutes each call was, he even had recordings of each call.

Instantly, I realized the power of this kind of advertising. Not only were publishers providing immense value to the buyer, but they also had the ability to dive deep into the mindset of the consumer making the call.

Publishers could understand what callers needed, why they called in, what questions they had, where they came from, and which ads were working best. This was data I would have killed for in any other kind of affiliate campaign!

My head buzzed with excitement. It was like I had unlocked the cheat codes on a whole new level of advertising that no one knew about.

When you understand the questions the consumer is asking the salesperson, you can start to piece together the puzzle of how to find more of those people asking the exact same questions. It's one of the absolute best ways to do market research because you're listening to actual sales conversations—with permission, of course.

In addition, those call recordings allow you to craft each individual ad to the exact mindset of the consumer, which means they're highly qualified when they call. And that makes for very happy buyers because they have an easier time closing more sales.

It may sound like an exaggeration, but when you do this right, it's like money just sort of rains from the sky. All because you're helping the consumer win, which in turn helps the buyer win, which means you win. I had found my new business model at last!

Here's how it works

I want to make sure you understand exactly how this works, so let's walk through a basic pay per call publisher flow from start to finish. Pay attention to the customer journey as we go through this process.

Step 0: Buyer learns about the benefits of pay per call campaigns

For a business owner to want to play in the pay per call arena, they have to know it exists. Sometimes they find out about it through a friend, colleague, or even a piece of online content somewhere. But sometimes they'll find out about it through you. The possibilities are endless when you take on the job of educating and attracting your own buyers.

But I'm getting ahead of myself. We'll talk in depth about how to find buyers later on. For now, let's just keep going with this example.

Step 1: Buyer creates a pay per call campaign

The buyer wants calls. They want traffic coming into their business, and they want that traffic converting into paying customers. It doesn't matter what they sell. Products and services can literally be anything from any industry anywhere in the world. The key component is that the buyer takes inbound phone calls to make the sale.

They sign up with an agency, network, or individual publisher that provides pay per call campaigns, and they get everything set up.

Step 2: The buyer promotes the campaign to publishers.

If the buyer is using a pay per call network or agency to build the campaign, that network or agency will promote the campaign to their publishers (also known as affiliates) to get savvy marketers on their team driving calls to the buyer.

If you brought that buyer on board on your own, then this step is skipped. You already know about it.

Pay per call networks or agencies can be helpful because they facilitate the relationships, the tracking, and the accounting between all of the parties. They sit in the middle and connect the buyers and the publishers together so that neither party has to go through the process of managing all those connections.

The upside of networks for publishers is you don't have to find and sign on the buyers. That's already done. You can simply sign up for a campaign and start doing what you do best. An additional upside is that networks generally offer faster payment terms than direct buyers, allowing you to recoup your advertising costs faster. That's important because you don't have to lay out as much cash before you get paid.

The downside is networks take a margin or a cut of that transaction between the buyer's payout and what the publisher or affiliate takes home at the end of the day. It's your choice whether you want to work with networks or not. You're in control.

Step 3: Publishers apply for the campaign

Because we're dealing with established businesses buying calls, the vetting process is more strict than other affiliate campaigns, and that's a good thing. It means the work you put into driving phone calls will pay dividends for a long time.

Once you're approved, you'll receive a unique tracking number that will trigger a commission for every qualifying inbound call you generate. In traditional affiliate campaigns, the publisher receives a tracking link. In pay per call, you receive a special phone number. Then you use that number in all your promotional materials. When someone calls that number, it's tracked and counted toward your commission.

You're going to hear this a lot throughout this book, but it's really important to understand that you need to use your own call tracking. A lot of publishers in this space don't because there's an additional cost to using a call tracking platform. But if you don't have your own data, you can't tell if the network's statistics are accurate. You will have zero knowledge of what's going on during that phone call because you won't have the recordings.

On the whole, pay per call networks are great. But mistakes and shenanigans can happen. Without your own way to verify calls and track data, you have no control over your revenue. There's simply no way to know for sure how many calls you're generating. Which means people can steal from you.

So, if you want to build a real business in pay per call, use your own call tracking to provide backup and keep everyone honest.

Step 4: Publishers generate inbound calls through their traffic sources

Now it's time for you to do what you do best—drive traffic! Publishers take their tracking numbers and create advertising campaigns in any way they like. As you'll soon see, pay per call is different from other online campaigns. You're not just driving clicks. You're driving calls. And that means you have a wider variety of traffic strategies available to you.

Sure, you can use the usual online techniques like paid advertising and organic search or social media. But other types of advertising work, too, like billboards, radio ads, and even classified ads in the local newspaper. I know that sounds old-fashioned, but your only goal is to reach people where they are at the exact moment they're in the most need.

If you're working on a campaign for a tow truck service, you need to reach people when they're broken down—which probably means mobile friendly geo-targeted paid ads. No one needs to read a blog about choosing the right tow truck company when they're stranded in a strange city. They just want help right away.

On the other hand, if you're working on a Medicare or insurance campaign, people are going to take a little more time to do their research before picking up the phone. That means you might want to build a blog and work the SEO to get it ranking for the right keywords. If it's an offer for a local agent, you might also consider a local radio or TV campaign or even writing articles for the local paper.

In my experience, marketers tend to gloss over the old-school methods in favor of fancy online techniques, which gives you an advantage over the competition if you can get out of your own head and into the heads of those callers. Who are they? Where are they? What information do they need before they call? And what strategies can you use that no other publisher is using?

One pay per caller I know does little more than place classified ads in local newspapers and periodicals every week with a headline and a phone number. He and his team of 14 people place thousands of these ads a week, and it's an eight-figure business. So don't dismiss an advertising strategy just because it's old school.

Once a consumer sees your unique tracking phone number, calls it, and stays on the line long enough to signal a legitimate prospect (instead of a wrong number) you generate a commission. Sometimes publishers use multiple numbers for each individual website or marketing channel they promote so they know exactly which blog post, ad, or article generated the call—it's split testing for pay per call.

I highly recommend this practice because it allows you to double down on the kind of ad creative or content topics that actually convert viewers into callers and drop the ideas that don't. This is another reason for having your own call-tracking platform. You want to be able to generate lots of your own phone numbers so you can see what channels and creatives are working for you.

Step 5: The call is tracked and, if it meets certain criteria, generates a commission

The final step is for the call to be routed through a tracking platform so it rings in the buyer's office. From there it's up to the buyer's sales team to provide information and close the sale. Whether they make a sale or not, you get your commission as long as the call reaches a certain criteria. The main criteria is a certain duration for the call. Every campaign is different, but the required duration is usually 90 seconds or so.

One thing to note is that there may be more than one buyer for any particular campaign. Imagine you're running ads for chiropractors in Manhattan. Not every office is going to be open at the same time, and what if all their sales people are busy—do you want that call you generated to just hang up because no one's answering? Of course not! You want every single call to be counted, not only because it's revenue on the line but also because there are real people in pain on that call. They need help immediately.

So tracking platforms will route and load balance calls between several different call buyers. There might be five or six home service companies taking calls, or maybe as many as 50 if it's an insurance campaign and each state is licensed separately.

It's not up to you to decide where the consumer gets help. The buyer's sales team needs to do the actual selling. Your only job is to drive the calls.

If you're working through a network or agency, they will credit you with the phone call and keep track of the commissions you're owed. They will typically use a third-party tracking platform like Ringba to provide transparency and ensure that statistics remain unaltered. We always recommend that publishers also track the calls in case there's any discrepancy or questions about how much money you're owed.

The problem with networks that use their own platforms or do the call routing themselves is there's no guarantee the statistics are accurate, and they never allow any outside company to audit their platforms. So when you're looking at any campaign to promote, ask how the call routing and tracking is done. If it's internal, you absolutely need to have your own tracking platform or you're at their mercy.

What counts as a "qualifying call"? Every campaign is going to have unique qualifications the callers have to meet. Those may include geographic location, certain demographic information, the duration of the call, IVR (interactive voice response) choices, and other criteria depending on what the call is for.

For instance, let's say you're running a campaign for debt consolidation, and one of the requirements is the caller has more than $10,000 in unsecured debt. If the caller answers an automated IVR choice saying they have less than the required amount, you probably aren't going to get paid for that call.

Or let's say you're running an insurance campaign for a company that's only licensed to take calls in California. If you send them a bunch of calls from Massachusetts, those calls are probably not going to qualify because they don't meet the criteria of the campaign.

Don't get freaked out by this. The criteria protect the buyer from paying too much and keep the buyer happily paying for those qualified calls. All you have to do is adjust your marketing to make sure the right people are picking up the phone. We'll cover all the ways to do that throughout the rest of this book.

Pay per call math

Thankfully, pay per call math is not complicated. Lord knows if I can do it, you can do it. I graduated high school with a C average, below most of my class, and I only managed to complete remedial math my senior year.

As embarrassing as that is, I'm proud of the fact that I can be a shining example of what's possible through hard work and dedication. You do not need to be a genius to understand how any of this works. But you must understand how the math works. No matter what business you're in, if you don't know how the math works, then you're probably going to lose money.

If you're a fan of the TV show *Shark Tank* like I am, then you know the number one thing the Sharks complain about is that an entrepreneur doesn't know their numbers. Go through this section as often as necessary until you're certain that you understand the math.

The first thing we're going to get comfortable with is counting the number of calls we generate. Sometimes people refer to calls as phone leads, leads, inbound calls, or raw calls. Regardless of what we label a phone call, we want to count every time a consumer picks up the phone and dials. This is how a pay per caller gets paid: per call.

Easy, right?

Next we need to understand what a duration-based conversion is. Some campaigns pay per *raw* phone call. That means every phone call that is tracked counts for a commission regardless of how long the consumer spends on the phone.

However, in most of the major categories, like insurance and financial services, buyers will only pay when a consumer stays on the phone for a certain minimum amount of time. We refer to this as the *conversion duration*. A *conversion* is the label we give to a call that meets all the requirements and qualifies for a commission.

While duration-based conversions do create a burden on the publisher to make sure consumers are interested and qualified, they also allow the advertiser to build long-term sustainable campaigns. If you're having trouble getting qualified calls, it's an opportunity to get creative and improve your marketing skills.

Calculating conversion rates

An important metric that all pay per callers track is the conversion rate of their calls. While software almost always determines this math for you, it's important to understand how to do the calculation yourself.

For instance, if you generate 100 inbound calls, and 25 of those calls meet the minimum conversion duration, then your conversion rate is 25%.

25 Conversions / 100 Raw Calls = 25% Conversion Rate

Now this is where things get a little interesting. If you have two buyers who are both selling the same product, they may pay out two different rates for the same calls. Buyer A may pay $20 on a 90-second duration, and buyer B may pay $25 on a 90-second duration.

Logic would dictate that you would want to send all of your phone calls to buyer B because they pay more money on the surface, but the reality is that we need to calculate the *revenue per call* or RPC of each buyer to see who is really paying the most money.

Let's assume that you send both buyer A and B 100 calls each. Buyer A has a conversion rate of 32%, and buyer B has a conversion rate of 24%. You would calculate the RPC as follows.

25 Conversions / 100 Calls=25% CONVERSION RATE

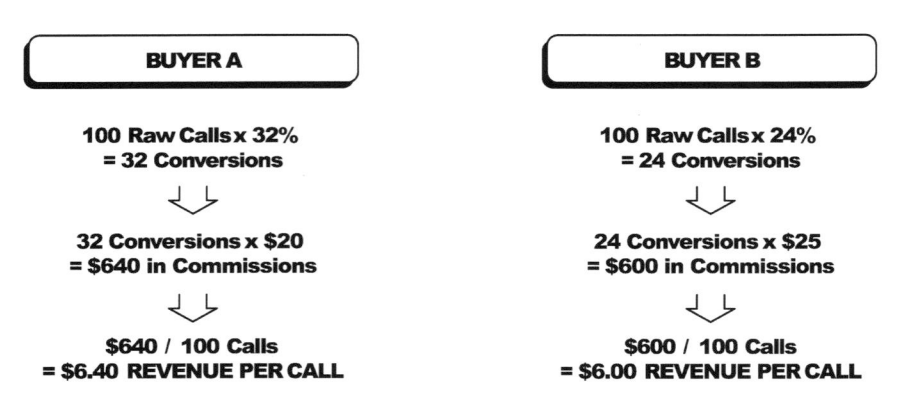

BUYER A

100 Raw Calls x 32%
= 32 Conversions

⬇

32 Conversions x $20
= $640 in Commissions

⬇

$640 / 100 Calls
= $6.40 REVENUE PER CALL

BUYER B

100 Raw Calls x 24%
= 24 Conversions

⬇

24 Conversions x $25
= $600 in Commissions

⬇

$600 / 100 Calls
= $6.00 REVENUE PER CALL

After reviewing the math we have determined that even though buyer B is paying a bigger commission, they're actually paying less money overall on a per call basis because their call center or advertiser converts those calls at a lower rate.

Lower conversion rates are usually due to hold times, poorly trained agents, too many new agents or trainees, agents cherry picking phone calls for whatever reason, more strict qualification processes over the phone, and various other reasons.

Regardless of the reason why, we always want to make sure that we're sending our calls to the buyers that are paying the most on a calculated revenue-per-call basis. *That* is how we maximize the amount of commissions we receive for every call we generate.

Let's do a story problem this time. You've set everything up, and you are generating inbound calls for dental offices in Houston, Texas. You have three buyers who are taking the calls.

In this scenario you can see that Dentist B is the winner and you would want to send them as many of your calls as possible to maximize your revenue per call. While in this story problem there isn't a huge difference in overall commissions, that's not always the case. If you don't stay on top of your campaigns and the numbers, you will definitely leave money on the table.

How Pay Per Call Works   25

DENTIST A
is paying $30 per call
on a 120-second duration
with a 19% conversion rate

DENTIST A

100 Raw Calls x 19% = 19 Conversions
⬇⬇
19 Conversions x $30 = $570 in Commissions
⬇⬇
$570 / 100 Calls = **$5.70 REVENUE PER CALL**

DENTIST B
is paying $25 per call
on a 90-second duration
with a 25% conversion rate

DENTIST B

100 Raw Calls x 25% = 25 Conversions
⬇⬇
25 Conversions x $25 = $625 in Commissions
⬇⬇
$625 / 100 Calls = **$6.25 REVENUE PER CALL**

DENTIST C
is paying $20 per call
on a 60-second duration
with a 30% conversion rate

DENTIST C

100 Raw Calls x 30% = 30 Conversions
⬇⬇
30 Conversions x $20 = $600 in Commissions
⬇⬇
$600 / 100 Calls = **$6.00 REVENUE PER CALL**

Understanding call flows

How calls are routed and balanced through a tracking system can get pretty complex, as you'll see later on. But right now it's important that you understand the basic call flows and how they work. You may end up using one of them, or several, depending on the campaign.

Direct buyer flow

If you're working with a direct buyer and not through a network, and you've got your call tracking all set up, this is how the simplest call flow works from a customer perspective. A consumer sees an advertisement, they call your tracking number, and it's routed to a buyer. The buyer's team answers the phone. If the criteria are met, the publisher gets a commission. Simple, right?

From a publisher or affiliate perspective, the publisher gets their tracking number from a software like Ringba, they create a mobile ad or whatever type of advertising medium they choose, and they put that tracking number on the ad. When an interested consumer calls that number, the software tracks the call and routes it to one of your buyers.

Then if the call meets all the criteria, the software tracks that, and the publisher gets paid directly from the buyer.

IVR input qualification flow

IVR stands for interactive voice response, which is the automated message you get at the beginning of a call that asks you to "enter your 16-digit card number" or "If you're calling about a new account, press 2." IVR is used to route people to the correct agents, and it's a big part of pay per call.

For this flow, the consumer sees an advertisement and calls the tracking number. They are then presented with an IVR and given a few choices. Whatever they enter into the IVR determines whether the call is qualified. Say a person is calling about debt relief, and we're asking if the caller has more than $10,000 in unsecured credit card debt. If they say *yes,* then they qualify. If they say *no,* they don't.

If the call is qualified, it's forwarded to a buyer, and the publisher gets a commission for that call.

You're not out of luck if they aren't qualified for the first buyer, because you can have multiple buyers for every offer. Maybe a disqualified caller for buyer 1 is actually qualified for buyer 2. So your software automatically routes the caller to buyer 2, and payout is issued.

Warm transfer flow

This flow is similar to IVR, but there's a human doing the qualifying. A consumer sees the ad, calls the tracking number that sends them to call center 1. The qualifying agent will ask a list of questions to see if the caller is a good fit for the buyer. If they are, then the first call center transfers the caller to a second call center. They might say, "OK, Mrs. Jones, you qualify for this special deal. Let me get you transferred over to Geico."

Typically in a warm transfer, the conversion event is still generally based on duration, but not always. When the specified conversion event is reached, the publisher receives their credit. Now, the qualifying call center needs to get paid as well. So maybe the affiliate gets $10 in commission, and the qualifying call center gets $10 in commission, but both of those events are recorded after the transfer.

Warm transfer via network flow

Whenever a network is involved, it's possible that lots of people have their hand in the money. There could be other networks or brokers involved. A warm transfer via a network typically looks like this: A consumer sees the advertisement, calls the publisher's tracking number. The network credits them if that call meets the requirements.

Now, what's happening here is the network has a deal with the other call centers to convert the callers into qualified transfers. So they're the ones that are making the majority of the money here, and the publishers driving the traffic are getting a very small piece of the action. In these situations, it's possible that the publisher may get paid even when the calls aren't qualified. It's also possible that the network ends up getting multiple payouts. It just depends on the campaign. But you can assume that the party making the most money, outside of the buyer of the call, is going to be the network, not the publisher.

Multiple brokers flow

A broker is anyone who's not the final buyer of the call. So an organized network is a broker, but someone who's just slinging calls and hustling connections is also a broker. You never really know how many people sit in the middle of these call flows unless you're actually working with the call center that buys the calls.

In this call flow, a consumer sees the advertisement, they call the publisher's tracking number, and a network or some other type of broker is going to credit the affiliate with a commission if the call meets the requirements.

The network then turns around and sells the call to another network or broker. Then broker number two accepts the call and credits the originating network with some type of commission. Network number two can then sell the call to either a final buyer or, believe it or not, even another broker or warm transfer center. There can be many brokers in these flows, so be careful about this when you start working with networks. The buyer may be paying out $85 for the call, but by the time that money trickles down to you, it's like 25 bucks.

The closer you can get to the buyers, the more opportunities you can create for yourself. And once you experience a diluted payout a few times, you may get really excited to learn how to find and manage your own buyers. Don't worry, I'll show you exactly how to do that in later chapters.

Why call tracking is such a big deal

"What do you mean you're not paying for any of these calls?" That's a terrible way to start off a conversation, and unfortunately for me, it went downhill from there.

When we first started experimenting with calls, we didn't have the proper tools for the job. (In our defense, they didn't exist.) Our network would give us tracking phone numbers for our campaigns, and we just trusted them to properly report and manage our call flow. As we got better, our volume grew, and that's when we realized that there were all sorts of problems happening.

With a click or a lead in a regular affiliate campaign, you can essentially send unlimited volume anytime you want without having to really think twice about it. If you have a campaign explode with clicks or leads, you need to have a decent tech guy or add more servers, but your campaign will be just fine.

Calls however, are an entirely different animal. The thing about a phone call is that you have to have a human available to answer that call *at the very moment in time* a consumer picks up the phone.

People hate waiting on hold. When they're ready to buy or they have questions, their patience disappears because they want their desires satisfied immediately. So if you are working with a single buyer like we were, and all your calls are going to the same place, you're at their mercy. If they don't answer the phone in a few seconds, the consumer is going to hang up.

If you're driving calls directly to a single buyer, you can only send them exactly as many calls as they have humans available to answer the calls. Makes sense, right?

Now, if that business is buying calls from you, it's highly likely that they're also buying calls from other publishers as well. If one of those publishers ramps up traffic unexpectedly and fills all the available agents, your calls will start to drop and your campaigns will quickly become unprofitable.

The same goes for the other side. If you're a buyer and you can't answer all the calls your partners are sending you, those partners will look elsewhere to sell their calls, and the first place they're going to go is your biggest competitor.

It doesn't matter where you are in the value chain, if your calls aren't getting answered you have a big problem!

Every time we started to find success and scale a campaign, we asked for as much data as possible from our buyers. Unfortunately for us, we wouldn't get it until weeks later, if we got it at all. By then, it was already too late, and we had usually lost a bunch of money.

The problems we were facing seemed insurmountable.

We had no idea which advertising source was actually producing calls, so we couldn't tell which was profitable or not.

We had no insight into how many calls were being answered or how many calls were converting into billable commissions.

We had no guarantee our calls were going to get sold, so we couldn't optimize our advertising campaigns.

We had no way to do quality assurance and listen to the calls to make sure consumers were happy or the buyers were doing their job correctly.

We had no idea if the buyers were putting people on hold or dropping the calls.

We had no recourse if the buyer's phones went down, didn't have agents available, or they bought too many calls from other publishers.

We had no way to deal with repeat callers who wanted to talk to someone else.

And we had no way to validate that the amount of money we were getting paid was correct.

It was an absolute nightmare!

No Tracking =
- No optimization
- No quality assurance
- No guarantee of payment
- No traffic oversight
- No buyer oversight
- No idea how many calls were answered
- No idea if you're dropping calls
- No guarantee calls get sold
- No recourse if buyers' phones go down
- No recourse if they get flooded with calls
- No payment for duplicate calls

We took all the lessons we learned and built the perfect tool designed for people who are generating, selling, buying, or trading their phone calls. This software solves all of the problems that we paid dearly for when we got started.

Get your own tracking numbers—Ringba will give you as many as you want. Otherwise, you will have no idea how much money you could potentially be losing, when you finally build a successful campaign.

The whole point behind learning the pay per call game is to have 100 percent control over your own destiny. When you have your own tracking system, you have power. If one network goes down or doesn't treat you right, you simply push a button and send those calls someplace else. You're in charge.

Now that you know how this whole operation works, you may be wondering which niches and verticals you should look into. That's coming up next, so keep reading!

CHAPTER THREE

THE SIMPLEST WAY TO GET STARTED

"Holy shit! They're paying you how much?"

I was amazed when I got the call from my good friend and client Anthony Sarandrea, the CEO and founder of Pocket Your Dollars. I remember my exact words to him: "I am so proud of you, bro!"

Almost five years earlier, Anthony and his team were building their call business and reached out for help. They needed a better platform to carry their business into the future. Like Harrison and me, Anthony was tired of the affiliate curse and wanted to build a business that had real enterprise value that he could one day sell.

We worked together to grow and optimize his business, and to find new verticals and partners to work with. Anthony and his team were always thinking about the future and always worked harder than anyone else to overdeliver for his customers. We've had many weekend and late-night calls riffing about the business and how to grow.

And then he tells me he was offered $50,000,000 in cash for an affiliate business! I had never seen anything like it in performance marketing, and outside of pay per call, we probably won't. When you're selling products and services that may disappear in a few months, no real company is ever going to acquire your business. But when you're delivering highly qualified customers who are ready to buy right now to giant public companies that have quarterly earnings to satisfy, man oh man, is pay per call an exciting opportunity.

What are you optimizing your business for? Do you value simplicity or do you love a complex challenge? Is your goal to build maximum enterprise value? Or maybe you're seeking time freedom and independence. No matter what kind of business you want to build, you can do it with pay per call. There is no single way to proceed here. I'm going to give you every tool and strategy I can think of to help you be successful. Just know you don't have to use them all, and you can start small and grow whenever you want to. There are lots of paths up the mountain. And, who knows? You may find a better way that I never thought of. (If you do, be sure to tell me. M'kay?)

With that said, let's walk through what I consider to be the simplest path to success in pay per call. It's not the only way to set up your business; it's just what I would do over the course of about a year to build a solid foundation.

Whenever I've moved into any industry, I start out as an affiliate so I can understand how that industry operates. The simplest way to do this with pay per call is to drive calls to networks using online advertising. Working with networks will teach you what to do and what not to do as well as build up your treasure chest of failures that you can use later when you're scaling your business.

Start with one network. Then, over time, you can create relationships with multiple networks and load balance your calls between them. You will need call-tracking software to do this, but it's the best way to remain in control of your destiny.

Once you've got a handle on multiple networks and everything is moving smoothly, you're going to want to go after direct buyers. Networks are basically brokers taking a cut. They're useful in the beginning because they've done all the business development work, so you can just sign up and get started. But the goal is driving traffic to direct buyers as soon as possible.

Now, don't destroy your network relationships. That's not smart. Instead, simply de-prioritize the networks in your tracking software as soon as you bring on some direct buyers that are paying you more money. Then, if your direct buyers are overloaded or unable to receive calls, you can still send the calls to your networks. That way, you're optimized to make as much money per call as possible.

Next, find more direct buyers and call centers, and sign them up to work with you. Reduce your dependence on the networks until they're just your backups. One exception to this is if the network has an exclusive campaign that you can't get anywhere else. That's actually valuable. But if the network is just brokering the calls—buying them from you and selling at a higher price to someone else—then you're better off finding your own buyers and getting higher payouts. The second exception for more advanced pay per call businesses is if the network has additional allocation to your direct buyers and you're capping out. For instance, an insurance company may give me $50,000 in capacity, and the network may have $250,000. So, after I fill my cap, I can keep selling through the network. (And don't worry, we'll talk more about maximizing capacity later on.)

The networks I work with don't like it when I give this advice, but I do it because I want them to work harder to get exclusive campaigns. That's really how a network builds its competitive advantage. And I want them to succeed as much as I want you to succeed.

Once you've removed your dependence on networks and have several direct buyers in place, you're going to run the same play and cut out the direct buyers as well. You create your own network of buyers, and that way you become vertically integrated. For example, if you're generating phone calls and selling to a small insurance agency, and then routing calls to hundreds of micro-buyers under their umbrella company, you're getting the highest margin per phone call you possibly can.

This is also the most amount of work you could ever do in pay per call. So if you don't enjoy a constant seven-days-a-week hustle, building your own self-serve buyer network may not be in the cards for you. Not everybody loves working as much as I do. I get that.

Finally, you're going to create enterprise value in one particular vertical so you can eventually sell your company. That's the end game. You're going to create your own brand, like 1-800-Dentist, so that all those calls flow through your brand and you create something of value that people can remember. You want people coming to you to do their comparison shopping so you can funnel them to the right buyers. The assets you create like blogs, YouTube channels, social media followings, podcasts, and the like, each contribute to that enterprise value and are a strong key to long-term business success.

I'm not going to lie. This is a lot of work. You're going to be building a team, building properties on the internet, developing your own websites, managing SEO, and creating content. And all that's in addition to driving traffic! I personally think it's worth it to basically own an entire niche and have all the traffic coming to me. I'm willing to put in a year's worth of sweat equity to see exponential growth on the backend.

Once you have a brand built, that's when you expand into TV, radio, and out-of-home advertising in certain markets to brand those campaigns and really get into people's heads. And you might select four or five verticals that are closely related so you can upsell and cross sell other products and services to clients.

The next level is bringing on affiliates of your own so you don't have to do all the traffic-generation work alone. Affiliates are really smart. If you know that your team can generate a phone call for $10, you can pay affiliates $8 and see what they come up with. You move your cost of acquisition down while opening up an opportunity to scale and an opportunity for another affiliate to build a business. That way, you're basically providing them the same opportunity that you started with at the very beginning.

How pay per call networks function

Since you're most likely going to start by working with a network, let's take a closer look at how they function. Pay per call networks exist to make life easier for affiliates. They do all the business development, finding buyers, setting up the systems, and, in theory, reducing everybody's risk. That means they spend a lot of their time finding multiple buyers and sellers of calls to gain as much coverage and capacity as possible. They then use technology and staffing to manage that entire business on both ends of the value chain. Networks are in the middle. They manage multiple buyers and multiple publishers, and make sure everyone gets paid appropriately. They have two teams of people doing all that work—a publisher team that works with affiliates, and an advertiser team that works with the buyers.

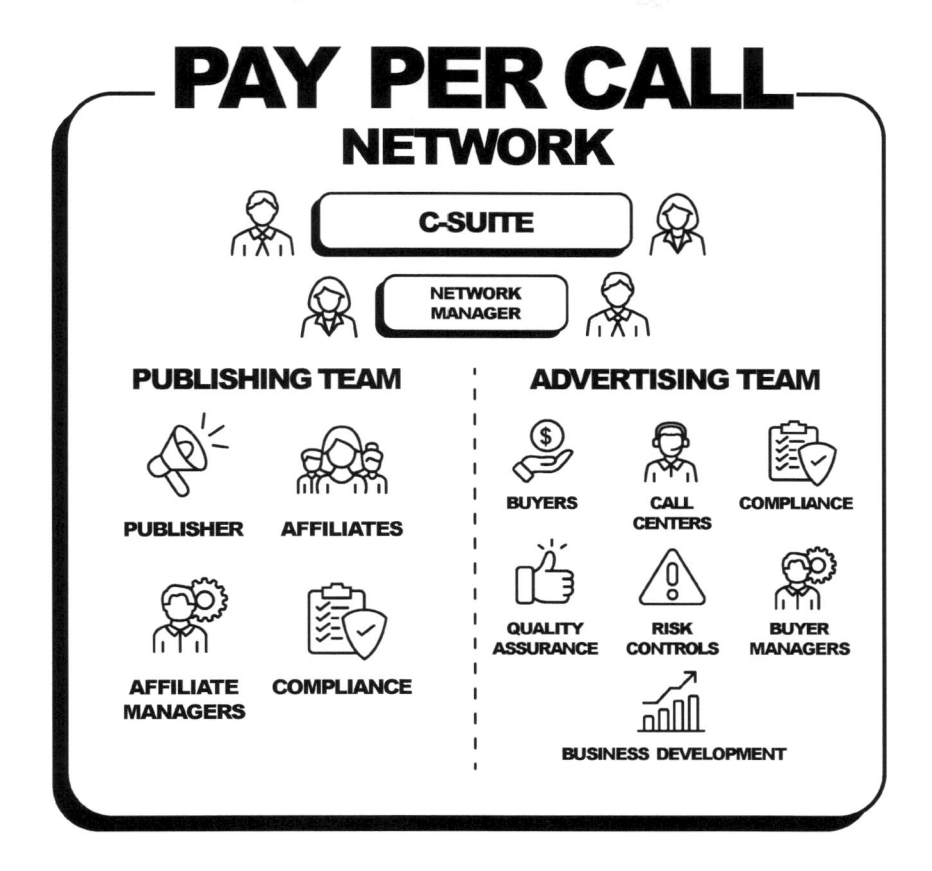

PAY PER CALL NETWORK

- C-SUITE
- NETWORK MANAGER

PUBLISHING TEAM
- PUBLISHER
- AFFILIATES
- AFFILIATE MANAGERS
- COMPLIANCE

ADVERTISING TEAM
- BUYERS
- CALL CENTERS
- COMPLIANCE
- QUALITY ASSURANCE
- RISK CONTROLS
- BUYER MANAGERS
- BUSINESS DEVELOPMENT

Depending on the network's size, there will be different people you'll be dealing with, the most important being your affiliate manager. Networks are in competition with each other, as well as other media companies and agencies who all want your call traffic, so the affiliate manager's job is to build relationships with the publishers and make sure the calls keep flowing. These people have all the inside information, tips, tricks, and whatever you need to grow your business. They often even have the power to give you a little payout bump. It's literally their job to help you succeed.

So, do yourself a favor and keep a great relationship with your affiliate managers.

Publishing teams at bigger networks will also have quality assurance (QA) and compliance people. They are listening to the phone calls and making sure that the traffic sources and landing pages all follow the rules of a campaign. They're checking up on publishers to preserve the integrity of the campaign. You won't see this at smaller networks, although good affiliate managers will try to do that job while the network is growing. But there will almost always be issues if a network doesn't have someone doing QA and compliance, because affiliate managers cannot effectively manage publishers, grow the business, *and* do compliance. Those are all separate jobs.

I'm not saying don't work with smaller networks. Just be aware that shenanigans happen—buyers play games with the rules, networks take advantage of publishers. It's a red flag if there's not a dedicated compliance officer. Someone has to play referee.

Every industry has its own rules you have to play by. Usually those rules are made up by the government or regulatory commissions that tell the buyers what they can and can't say in their marketing. If a publisher is sending traffic and breaks those rules, the *buyer* could be punished because "they should have known" what was happening. And sometimes that means fines and sanctions in the millions of dollars. These are big businesses in highly regulated industries. If you can manage to work with the rules and send high-quality traffic to these buyers, you're going to do really well.

And honestly, if you don't like the rules, that's just too bad. You won't be allowed to play. Networks deal with lots of publishers, and buyers may be harder to come by for certain types of campaigns. They are very concerned with compliance, because if a buyer catches a publisher doing something wrong, it can damage the whole network relationship. And, sorry, but affiliates are easier to come by than buyers. The rules are really important. Read your contract and know what you're signing on for.

The publisher team is also concerned about volume consistency. Call centers live and die by consistency because they have to preschedule labor. They have to preschedule people to come in to actually answer the phone calls. If you are fluctuating your call flow constantly, networks aren't going to like that because it's hard for them to predict and work with their buyers to get the capacity correct.

Communication is extremely important. I cannot stress this enough. Just starting a campaign and sending traffic without talking to a publisher manager first is a recipe for disaster. These people are responsible for your payouts. So if you want your money on time and you want to get payout bumps, communicate with them. Make them want to help you succeed.

We'll get into negotiating in a later chapter. But just to be clear, you should be negotiating the financials in this space every chance you get. Every change that happens should trigger you to ask for more. Quality goes up, get more money. Quality goes down, ask for more money! You should just be negotiating everything at all times. You can't do that if you don't have a great relationship with your manager. Got it? Good!

On the other side of the digital hallway, you have the advertiser team. These folks work with the call buyers, call centers, other brokers, and networks. Their job is to build relationships, sign up more buyers, and manage the backend of the network. They shouldn't be working with publishers like you. It's another red flag if the affiliate manager is also managing advertisers. They are two separate jobs, and one person really shouldn't be trying to do both. If you find a network where this is happening, it could be a sign that their buyers are lower quality or that they are working with numerous brokers on a single campaign. They're a middleman with lots of other middlemen, which lowers your payout.

Remember, you are in control here. When you understand how the system works, you get to choose which networks you associate with.

The advertising team should also have its own quality assurance and compliance people keeping tabs on the buyers and call centers to make sure there's nothing fraudulent happening on that side. Are the agents well trained? Or are they burning phone calls because they don't know what they're doing? Are they hanging up on perfectly good calls so they don't have to pay for the lead? The network's job is to keep everyone on the up and up.

In any sizable network, there will be plenty of other people running the rest of the company. The controller, accounts receivable and payable, management, C-suite executives—all the roles any larger company would have. You probably won't deal with these people as a publisher. In fact, you may never speak to anyone but your affiliate manager, and that's perfectly fine. If you go to trade shows, though, I highly recommend you meet the network managers and C-level people, and create a relationship with them. In time, you may have ideas for the network or you may have a special request. Higher-level people will have more power to make those things happen. Those relationships are invaluable and literally how I build most of my businesses.

Pros and cons of working with a pay per call network

As with everything in life, there's good and bad when dealing with networks. On the plus side, networks typically pay faster than direct buyers. You don't have to bill anyone, deal with buyers, chase down payments, or even worry about financial risk. If the network is playing by the rules, and you have your own tracking software to make sure they are, then you should just get paid on schedule every time.

They also offer a wider campaign selection and greater geographic coverage. A single call center or single buyer can only do so much. But with a network, you have 25 employees all working to create campaigns and attract buyers for you.

On the negative side, it's harder to get into a network if you're new in the space. And it should be, frankly. This isn't like a regular affiliate gig. There are humans involved, so it's more exclusionary, and you have to be on top of your game to be accepted. But if you communicate well and put in some effort to show up as a high-quality publisher, you'll be on the right track. And we'll go into exactly what that means a little later. I want networks tripping over themselves to sign you up.

Another downside is you're going to be working with lower margins than if you had direct buyers. Networks will be taking 35% to 100% margins. If they have direct buyers, it could be much higher. The networks are getting paid the most, but they're also doing the most work and taking the most risk. The margin problem is definitely a reason to start going after your own buyers as soon as you can.

Also, there's zero transparency. You don't know who the buyers are. You don't know what their capacity is. And if you ask your manager, they may or may not tell you. If the buyer isn't doing a good job, you can't work with them to train their agents better. You just have no visibility into the campaign whatsoever.

If the network isn't doing a good job, you have no control over that either. You have no control over the quality assurance process. You don't know if the network's really paying attention to the buyer to make sure that you get paid for every phone call. They may say they are, but you don't actually know unless you're tracking and listening to your own calls.

There's also going to be some favoritism. Your calls may be higher quality, but if an affiliate manager likes someone else more than you, they may not give you the same capacity. You're also going to have little recourse over disputes, especially if you do not use your own call tracking to keep them honest.

Now, there's also potential for mismanagement of funds in the mix. You don't know how the financial system at the network operates. You don't know if buyers are delinquent. You don't know the current cash standing of a network and if those networks pay weekly or if a network pays affiliates before they get paid from a buyer. Smaller networks tend to have more issues with this than larger ones.

This sounds like a lot of negatives, and I don't want to sway you away from networks because they're pretty much the best place to start. I just want to make sure you're going in with your eyes open so there are no surprises later. Pay attention to how they do business and know who you're working with. That's just good business.

Getting started with pay per call networks

Just because a network has multiple buyers, large geographic coverage, and purports to have plenty of capacity, that does not mean they are going to have 100% coverage and all your calls will be answered. On average, 25% of all calls anywhere in the world are abandoned due to lack of capacity.

Say a network is buying calls from you, and you happen to notice that in South Carolina, they drop 80% of the phone calls. That means they only have 20% coverage in that state. So *technically* they may have nationwide coverage, but that doesn't get baby new shoes when your calls aren't getting answered. You need to pay attention to these pockets and zones, especially when a campaign has geographic restrictions, to make sure that your buyers have the available capacity to actually take the call. It's an unfair situation, but because call flow is unpredictable, there isn't an easy way to solve the problem, so you simply need to be aware of it and build your business process appropriately.

It's in the network's best interest to try and keep their buyers happy. And since the network does not have to pay for a call that doesn't get answered, they would prefer to send a lot of traffic and have a few calls dropped than risk having those call center agents sitting idle, creating significant overhead, and lowering productivity for the buyers.

Call centers and buyers always want to have as little idle time as possible for two reasons. One, they are paying for those people, and, two, when call centers have a lot of action, it's jamming in there. Morale goes up. People get excited. It's like a team sport. They're literally high-fiving, ringing bells, having a great time. And the whole floor's productivity goes up when everyone is in a good mood, which means they're making more money.

If you're sending calls to a network and another affiliate ramps up their flow, your calls may start dropping because the buyer doesn't have enough capacity to handle the spike. That's the ideal situation for a network because it means they're going to make the most amount of money and keep all their buyers happy.

It's important to understand that networks will not reimburse you for advertising costs just because they don't have the capacity to handle your calls.

And most of the time, they are not going to notify you that they are overflowing, because it's not in their best interest for you to be out looking for more buyers.

If they know you're not using your own call tracking, they really have no incentive to tell you anything at all. For them, your ignorance is their bliss. You could be sending a steady stream of perfectly legit calls and not be getting paid for them, and you'd have no idea. You might think there's something wrong with your advertising. Or you might just decide this whole pay per call thing doesn't work, when in reality, the network is looking out for itself. They maximized their yield, maximized buyer capacity, and made the most money they could possibly make. They are not concerned that some poor affiliates lost out on commissions, because chances are slim that they even know about it.

Fortunately, you're smart and you know there are things you can do to maximize your own income.

Optimizing your call flow

As I mentioned at the beginning of this chapter, your goal is to start signing up your own buyers as soon as possible. With your own tracking software like Ringba, you can load balance your call flow across whatever criteria makes *you* the most money.

The buyer doesn't have to be some big call center, it can be just some guy named Jim. He might be a one-man show and can only take one call at a time, but that's OK. You want to send as many calls as possible to Jim because he's paying the most money. In the ideal scenario, Jim's phone rings off the hook and he loves it, because all day, every day people are like, "Hey I wanna buy your service." Then any overflow calls you have can be routed to your networks, starting with the ones that pay the most.

The growth path looks like this: You start with one network. Then you go get three networks. Then you find Jim and a couple more buyers like him. Then you build out your call flow exactly the same as a network would with your own technology. Then if the network doesn't want to pay you more money or doesn't want to negotiate with you, at the click of a button you can redirect your entire call flow to a bunch of other buyers.

Then you're in the power position in full control of your destiny!

Let's say you're just getting started, and you're going to work with two networks. In this example, network A is claiming they can handle unlimited call capacity 24/7. Right off the bat, if anyone tells you that, it's complete bullshit. It's not humanly possible. Anyone who claims to take unlimited calls isn't giving you the full story. I know I've said this a lot, and I'm going to repeat it again: Humans have to be available to take every call or the callers will hang up. They're not going to sit on hold for 26 minutes to get a new insurance policy. You're lucky if they will sit on hold for 26 seconds, OK? So you want those calls answered as fast as humanly possible.

Now, network B, on the other hand, has given you a cap of 100 calls per day to fill without providing any concurrency limits. This is also a problem because they are not telling you how many simultaneous calls their buyers can handle. That's what concurrency means, and it's a big problem. It doesn't matter to them, but it should matter to you. Because if two people call from your ads at the same time, you might get both answered, you might get one, or none at all.

To recap, network A offers unlimited capacity, which is nonsense. Network B at least, gave you a cap, but they didn't give you a concurrency cap. They didn't tell you how many calls you can run concurrently, so you need to dig that information out of them so that you can properly balance your calls.

Let's assume both networks claim to have nationwide coverage and will pay out after the call length has reached 90 seconds. Remember, just because they claim to have nationwide coverage, doesn't mean that that nationwide coverage is even. If one of those areas, like Florida, gets maxed out or is always on the phone, your Florida calls may bounce at a higher rate than your other phone calls.

This is another reason why you need sophisticated call tracking that can geographically load balance your calls for you. Then you can say, "This network is good. They pay me quickly. I want to work with them, but on this campaign they suck in Florida for whatever reason. Their buyer drops 80% of the calls. I wanna reroute my Florida calls somewhere else so that I can maximize the yield on that campaign."

Think about call-flow optimization like a jigsaw puzzle. You can't put a puzzle piece into the wrong spot. It just doesn't fit. It's the same with a call. If humans are on the phone, you can't park it at that buyer. You have to route it someplace else. So you fill in your puzzle with networks and buyers to make sure all the pieces work, and you get the most money out of your call flow.

And I'll be real with you, most affiliates are terrible at this. When you really get going and have complex call flows, you don't want to be figuring it all out with graph paper and bubble gum. You need technology that can figure it out for you at lightning speed.

So let's assume you were able to generate those 100 calls and use your call tracking platform to load balance the calls evenly to both networks. You split them 50/50 just to see what happens. You sent 50 calls to each but for some reason, network A only paid you for eight of those calls, whereas network B paid you for 19 of them.

That's really odd, you think. But there's a reason for it. Network A didn't have the capacity available to take those calls, even though they told you they did. What are you going to do when you notice the discrepancy? Reroute the calls to network B! You are in control.

How does it feel to know you can pick and choose where to send *your* traffic?

I know you've been used to an abusive system where the rug can get pulled out from under you at any time. But that's not how you operate anymore. What you build has value. You own it. And you can sell it to whomever you want for as much as you want. You're not at the mercy of fly-by-night operators and sketchy offers anymore.

I hope that gets you excited about the possibilities for this space.

Are you ready to start picking verticals and looking for your first offer? That's coming up next. Keep reading!

CHAPTER FOUR

CHOOSING VERTICALS

Your first step in this business is figuring out what industry vertical you want to play in. I started my pay per call journey promoting merchant accounts because I knew the industry really well. I used to bang on doors in downtown Detroit selling them to businesses. I was so excited to get started and figure this out on my own with a big direct buyer that I skipped the networks and negotiated a campaign directly with the CEO of a large merchant account purveyor.

It was all real casual. I was just going to send them some calls—no big deal. The CEO didn't really tell anyone on his call center team what I was doing because it was just a few calls.

Now, this was long before we built Ringba. I had no IVR, and no way of tracking the calls. I knew nothing about quality assurance. I really had no idea what I was doing. All I knew was this company had deep pockets and I was going to drive a ton of calls their way. So I tried a bunch of unique angles inside AdWords, and one of them hit big. People phoned the call center in droves. Unfortunately, none of them were qualified.

One of my Google keywords got expanded because I didn't use exact match phrases, and I had no negative keywords. So, people looking for customer service related to a certain type of credit card terminal were routed to the sales call center. It was a disaster! No one could figure out why hundreds of people were calling sales and asking for customer service.

Well, eventually someone figured it out and called me screaming bloody murder. I had flooded their call center. The sales agents had transferred calls to customer service, and that department got overwhelmed by callers who weren't even their customers. I basically destroyed a massive call center for an entire day, probably costing them tens of thousands of dollars. Total fiasco.

Needless to say, I didn't get paid, and that company probably never worked in pay per call again. But there was one positive thing I took away from that fiasco: I did make the right call going into that vertical. People really responded because I knew the consumer intent. I knew all the questions the customers had and how to trigger their emotions and curiosity. I knew the economics of the advertiser. All that knowledge allowed me to create the right ads, the right landing pages, and to come up with some great keywords. It was just an easy campaign to get up and running.

Campaigns need to be thoughtful and controlled. There's a delicate balance between how many calls you send and the buyer's capacity. I should have been much more conservative, but I was definitely in the right vertical.

Choosing a vertical

The best way to think about what's possible with pay per call is to think about what types of businesses take phone calls from people who are interested in their products and services. It can get overwhelming when you realize that covers just about every industry and sector. But I want to say up front that just because a vertical, industry, or niche isn't mentioned in this book doesn't mean pay per call programs don't exist for them or can't be created. As long as you have the technology right, you can build out your own offers and buyer networks fairly easily. Right now, though, we're going to focus on large industry segments that already exist and have a robust ecosystem already operating today. Here's a big list of those:

- Insurance
- Home Improvement
- Home Services
- Senior Care
- Legal Services
- Travel & Tourism
- Health & Medical
- Education & Training
- Financial Services
- Auto & Automotive

The first thing you want to look at is what's available, what has capacity, and what's popular. The easiest way to do that is to join all the pay per call groups you can find on LinkedIn, Facebook, and the private forums like Pay Per Callers. Poke around and see what industries are mentioned. In the beginning, you're looking for an industry proven to have high volume because it'll be the easiest to get up and running.

Now the highest-volume offers may be saturated, so you might also want to look at the mid-level offers that have plenty of capacity but aren't full of competition. I know people who are doing $100 a day, $1,000 a day, $10,000 a day, $100,000—I've even seen $1,000,000 a day in pay per call on a single offer. We have clients that are doing all of those. I know for a fact that pay per call volume can be huge, it's just a matter of figuring out where the volume is and then creating competitive advantage in that space.

The simplest place to begin is wherever you have a familiarity with the industry. If you have a basic understanding of caller intent, it's easier to create campaigns around that intent. You also have an advantage in understanding how the buyers build their business, so you can find direct buyers and have conversations with them. If you used to sell insurance, it just makes sense to look at insurance offers. If you used to be a plumber or contractor, home services might be a great fit.

Don't overthink this. What industry do you already know? Where can you speak the language and find opportunities that others might miss?

Now, if you absolutely hate the last industry you were in, maybe it's not the best choice. If you're going to turn this into a full-time sustainable business, then you need to enjoy what you're working on. Find a niche you can get passionate about, or at least one where you won't get bored researching and iterating hundreds of ads.

At this stage, research is really important. I know you're in a hurry to get the money rolling in, but this is critical groundwork. You need to understand what the customers want, and then you need to understand the geographic area of these campaigns. Is it the Pittsburgh slab installation or is it a nationwide insurance campaign? Will you be regional or can you take it all over the world?

Do your research and do it on mobile! This is important. Probably 99.9% of your calls will come from mobile devices. Follow the rabbit holes until you're certain you know who the target audience is, what they want, what their pains and problems are, what they ask about, and what all your competitors are doing.

You want to have lots of room to grow in whatever vertical you end up choosing because you're going to stick with one niche until you're making significant money with more than one buyer in your routing plan. The more you learn about an industry, the better you'll do with pay per call. The money is directly proportional to who you know, what you know, and how hard you work at it.

Working with pay per call networks

If you're going to work with networks to begin with, you first have to be accepted by one. I cannot stress this enough, you must show up at your best. How you approach a potential network matters. Be professional. They need publishers. They need you. They really want to work with as many great affiliates as they can. But they have no time for people who can't communicate well. There's too much at stake to work with chuckleheads.

You're looking to set a fantastic first impression and then build a long-term relationship. So when you first approach a network, let them know a bit about your background in good, clean English. Use a spelling and grammar checker, if you're not great at communicating in English. You can even use AI to help you generate a polite introduction email.

I wasn't great in school, so I had to learn how to present myself as a professional. The best trick I learned was to read every email or communication out loud word for word. It's amazing how many spelling mistakes and other errors you'll find when you do that. Triple check yourself before sending introductions.

✓ DO THIS

✉

To: Jeff@acmeppcnetwork.com
From: adam@paypercallrevolution.com

Subject: Adam Young - Digital Marketer

Hi. My name is Adam, and I'm getting into the pay per call space. I'm a digital marketer with a background in the plumbing and home services industry. My family has had businesses in this space, so I know it really well. I think I can add a lot of value to your network in home services. I'm starting to drive calls in this space and would be interested in learning more about your company.

NOT THIS

✉
To: Jeff@acmeppcnetwork.com
From: mafiaguy227@hotmail.com

Subject: Hey!

hey bro . work with me why you not approving me ?

You'd be surprised at what lands in a network's inbox!

Understand that these networks want to make money. They're not going to waste their time with you if they don't think you can help them with their goals. If you don't present yourself as a legitimate player, why should they take you seriously?

To create a real relationship with the people behind the network, reach out to them, ask them about their business. What's important to them? How did they get started in the industry? What kind of people do they love working with? Be genuinely curious about them. That's how you make friends. Reach out to lots of people at the network, not just one. You can find them on LinkedIn or other social media platforms. See if you can develop a good rapport with a few different people.

An approach like that tells them you're serious, that you want to learn about the space and the best way to work with them. If they know that you care, they're probably going to work with you or at least give you a shot. That's all you're looking for if you're new. You're looking for a shot. That shot is not necessarily going to come easy, OK? These people don't owe you anything.

Next, make it easy to contact you. Set up an email signature with your actual name. Add your phone number, social media accounts, and all your contact information. If you don't make it easy, they won't bother responding.

Before they reach back out to you, there's a good chance they'll look you up on Google or LinkedIn or Facebook or wherever. What will they find there? Is your online presence as polite and professional as the email you sent?

Do a little editing on your social profiles. Make sure there's a photo of your smiling face as the profile picture. Write a professional sounding bio. Get rid of any compromising pictures or posts that might make someone decide not to work with you. It's great if you have your own website, but that's not critical. At the very least, have a professional profile set up on LinkedIn.

If you've been in the pay per call space for a while and you're looking to add additional networks, you might write something like this:

> Hi Bill,
>
> My name is Adam Young with Call Experts, and we've been in the digital marketing space for three years. We've been seeing a ton of success in the plumbing vertical, currently generating 30 high-quality calls per day.
>
> The biggest problem we have right now is finding more buyers. I hear great things about your network and would love to work together. Ultimately, our goal is to find long-term partners to work with so that we can dramatically increase our volume.
>
> I'd like to better understand how your company does business and how we might work together. Are you available sometime this week for a quick call?
>
> Looking forward to meeting you in person,
>
> Adam Young
>
> Phone Number
>
> Website and LinkedIn URLs

I'm assuming we're going to set up a call because I take myself seriously. There's absolutely no reason for them not to get back to me. And that's what you need to do when you're reaching out to anyone in this business.

Besides reaching out to networks directly, there are lots of other places you can start making relationships and finding great offers to work with. Just don't forget to present yourself professionally and proofread your messages, especially if English is not your first language. Here are my favorites places to start looking for offers:

PayPerCallers.com This is Ringba's private networking group. I'm in there, my team is there, and thousands of other pay per callers in all sorts of niches and industries. It's a tight-knit community. And, if I do say so myself, it's really the best place to start learning about pay per call and diving into actually driving calls.

One of the best things you can do for yourself is to dive in, make friends, and start building a reputation for yourself. Whenever you can help someone out, do it. Teach something you know, even if you're new. Add value. Ask questions. And always be learning. Introduce yourself, tell us where you're from and what you like about marketing. Let us know what verticals you're interested in. Before you know it, people will be reaching out to you, making recommendations, and you'll be on your way.

There are buyers, sellers, and networks on the platform. So you can post about what kinds of calls you want to work with, too. People will reach out to you with opportunities.

Facebook Groups There are lots of Facebook groups for pay per call, and you should join them all. Ringba owns the largest one, and you can consider this your personal invitation to join us. All we ask is that you answer a few simple questions so we can keep the spammers out. We moderate the group and don't tolerate bad behavior, so be respectful of your fellow pay per callers.

These groups are geared toward buyers and sellers of calls. So, almost all of the networks are in our groups as well, including the account managers, affiliates, and brokers. There are thousands of people in our group, and while they may not all communicate regularly, they're all reading the posts.

Facebook groups are a great place to find new contacts and business partners, share your journey, communicate and ask questions. The biggest experts in pay per call are there to help you out. The most successful people communicate clearly and proofread their posts. They take themselves seriously, so others do, too.

LinkedIn Groups These are less active, but they can be a great resource because so many people conduct business on that platform. You can make great connections with potential partners and find buyers and sellers just by searching. And you can reach out to them all directly through the groups, or with private in-mail if you have the paid premium version of LinkedIn. Everything you need is right at your fingertips. You gotta love the internet!

Just like every interaction you have anywhere online, the best way to interact is to add value. Tell the person you can help them in some way. It doesn't take much to get a reply. Use the examples I showed you earlier. Don't spam people, and don't send incoherent messages.

Before you send any messages, make sure your personal LinkedIn profile is up to date. At bare minimum, it should show the current company that you're using to promote pay per call offers. Write a little about yourself and the industry verticals you work in, maybe the school you graduated from. And make sure you have a real picture of yourself with your real name. People will take you seriously if you're real with them.

OfferVault This website is another great place to find offers and networks to work with. All you have to do is search for specific verticals, find offers that interest you, and reach out. Tell them what you're interested in specifically like this:

"Hey, I saw that you had a final expense life insurance offer on OfferVault that pays $26. Is that offer still valid? I would like to work with you on it. I have final expense calls to send you. What are the terms of your offer and the rules of this campaign? I'd like to see if my traffic works with your offers."

It doesn't take much to get a reply as long as you're clear and professional.

Trade Shows Some people think industry events and trade shows are too much work. They'd rather stay at home tucked behind their computers. But let me tell you, trade shows are awesome! They are my absolute favorite way to do business in any space. We spend a ton of time and money attending trade shows, and it always pays off. I've met many of my friends and business partners at these events. My partner and co-founder of Ringba? Met him at a trade show. Just go to one, and you'll be hooked. They're so much fun!

Right off the bat, when you show up at a show, people take you seriously, even if you're a beginner. I remember a gentleman who came up to me at Affiliate Summit West. He had seen my *Pay Per Caller* show and said, "Hey, man. I saw the *Pay Per Caller* show, and you keep telling people to show up to trade shows, so I came. I don't know a whole lot about pay per call, but I would really like to work with XYZ Marketing, and I'm super excited about being in the space."

So, I spent 15 minutes talking to the guy. Then my team members spent another 20 minutes giving him tips and tricks on how to get started. Then someone from XYZ walked by, and we were able to introduce him to a manager in person. Bam! Relationship started. Opportunity created. It's that simple when two humans show up and shake hands.

You can build your entire business from a trade show. I learned just about everything I know about digital marketing at them. My business partner Harrison started in the space when he was just 13 years old. They wouldn't even let him in the door because he wasn't 18. He bought a pass anyway, parked himself just outside the door, and proceeded to make relationships that created millions of dollars in business for him as a teenager. All from just standing in the lobby. If he can do it, you can do it, but you have to show up.

There's no magic trick to mingling and networking at trade shows. You can't go wrong by opening with a compliment. *Hey, I love that tie.* Or

You've got a great vibe at this booth. At the very least, they'll say thank you, and that's a conversation opener.

People love to talk about themselves, and so all you have to do is ask questions to get the conversation started. Believe it or not, my favorite icebreaker at a trade show is, "So, tell me about what you do." Yes, that's extremely boring and basic, but it's also really effective. And if they actually like what they do, they're gonna talk to you about it.

So it's your first trade show, and you don't know anyone. What do you do? You walk up to someone, maybe you notice it says "Bill" on his name tag, and you say, "Hey, Bill! I see you work at the super amazing gift card company. Tell me about what you do there."

I also like to ask, "How long have you been in the industry?" That question gives me insight into whether that person is a veteran or someone who's new. And I always like to build relationships with the people who've been around forever because they know everybody in the industry and can connect you with great opportunities if they like you.

Another question I like to ask is "What are you working on right now?" I want to understand what their goals are, what direction they're moving in, and what their challenges are. Maybe I can even help them out somehow.

Now, how you present yourself in person is even more important than how you come across online. Dress professionally, even if others don't. Be respectful and friendly. And know what you want to accomplish at the event. Are you there to learn or to network? Who do you want to meet? What do you want to know? Trade show booths are some of the very best places to get an education on what works and what doesn't. These people know it all and love to talk to interested people about what they do.

Business cards may seem old fashioned, but they're really handy. You want to make it easy for people to stay in touch. Even better than cards is having your phone set up to transmit a virtual contact into their phone. I would have both available, just in case. It's more important for you to *get* the cards or contact information of people you want to follow up with than it is to give out a bunch of your cards. Most people will never follow up with you, but *you* are more motivated than that. You will follow up, right? (Nod your head *yes!)*

You're there to meet people, so don't waste any time. Talk to as many people as you can. Connect with them on LinkedIn, and start building relationships. By the time the event is over, you should be exhausted from meeting so many people. It's exhilarating and fun, and it will have a profound effect on your future.

How to evaluate an offer

How do you evaluate an offer to decide if you should promote it or not? Well, the first thing to look at is the payout and terms of the offer. The payout may be $25, but it only converts at a call duration of two minutes. Two minutes might be really long or totally reasonable depending on what they're selling. Some buyers make the call duration longer so they can afford to raise the payout rates and get the best publishers on board. But in reality, that higher payout might not mean more money in your pocket if the duration is too long.

You have to look at comparable offers to figure out if this is a viable option for you. Not only do you want to know the average payouts in the market, but you also want to have several buyers on a campaign. If you're running a really successful campaign and one buyer goes down, or they run out of capacity, or their call center explodes, or their tech stops working, or for whatever reason they can't take your call, then you're driving phone calls into dead air or a busy signal. You may not even notice it, but you're still buying the media. You can run into some pretty big losses that way.

Once you have a few comparable offers, look at the hours of operation. Do they take calls seven days a week or just on weekdays? When are they open? When do they close? If a call center is only open from 9 to 5 on Eastern Standard Time, then you probably need your ads to turn on at 8:45 in the morning and then stop at 4:30 in the afternoon.

You also want detailed information about the buyer's capacity. Lots of affiliates are happy to know they can send 200 calls per day. But you really want to know the hourly cap, daily cap, and your concurrency cap. And you want that on a day-to-day basis. If they're open seven days a week, chances are they'll have reduced hours or reduced staffing over the weekend. You need to know that so you can properly cap your media spend on an hourly basis to avoid bottlenecking your calls. You don't want spikes in traffic unless you've planned for them.

To find comparable offers, check all the pay per call groups you can find on Facebook, LinkedIn, and in private groups. Find as many comparable offers as you can. Talk to the account managers and find out the average conversion rates (for the publisher, not the buyer). Then look at the offers side by side and take your best educated guess as to what the average revenue per call is likely to be. Also look at capacity and concurrency to see where the opportunities are to scale a campaign.

If ad costs are going to run at least $15 per call, then you need a decent margin above that for the offer to be worthwhile. In this case, $8 a call is only going to lose you money. Maybe you need to rethink your caller acquisition, or maybe you need to find better offers or go direct to buyers. Find out as much as you can about the economics of the whole offer before agreeing to run it.

Try to have at least two or three buyers so you can get a better feel for the actual revenue per call. The reality is that some buyers really outperform. And if you're only working with one, then you don't know if you have a great one or a dud.

One quick note about Spanish-speaking audiences

You can absolutely crush it with Spanish-language campaigns in the United States. They're super high margin, and not a lot of people compete in that space because they don't speak the language or they don't want to take the time to figure it out. If you go to OfferVault, for example, and search for Spanish-language campaigns in pay per call, you'll see plenty of offers. Spanish debt settlement, Spanish pest control, Spanish motor vehicle accident, Spanish tax debt settlement, Spanish health insurance, Spanish Spectrum, Time Warner Cable and Internet in Spanish—the list goes on and on.

Think about your targeting carefully. It makes sense to run Spanish campaigns in places like Florida, Texas, Arizona, and New Mexico. And you can target by language on Facebook, Google, TikTok, and other platforms, too.

A few tips for Spanish campaigns...

Don't use Google Translate. Hire native Spanish speakers to help write and create your advertisements. You can find inexpensive translators on Upwork and other outsourcing sites. You can just translate your English copy into Spanish, but if you want it to convert at a higher level, talk to someone who has an understanding of the Spanish market. They speak and think differently, they have different mindsets. And tapping into that cultural piece is a special skill.

Watch your ads carefully to make sure your targeting is correct and the ads are firing to the right people. Sometimes your Spanish ads will get shown to English speakers, which means your ad doesn't perform, and that negatively affects the algorithm. The platform thinks you have a crappy ad, but really the targeting is off.

The best way to approach it is to target languages in certain geographic regions with high Spanish-speaking populations. If that campaign goes well in, say, Texas, then you can spread it out to other areas. Just keep an eye on your analytics to make sure everything is working as expected.

CHAPTER FIVE

BUILDING A PROFESSIONAL BRAND

A while back, I sat in the audience at a mastermind event put on by my good friend Jason Akatiff, the CEO of A4D, one of the most well-known affiliate networks in the world. I had worked with Jason and his team for years running all different kinds of campaigns, but I had never been to a mastermind event before. Being from Michigan, though, I figured any excuse to head to San Diego in the winter is a good one.

Jason introduced the next speaker, a brand manager from one of his advertisers. I was expecting him to give us some insights on how to run traffic to his offers, or maybe the inside scoop on what affiliates were doing to warm up consumers to increase their conversion rates. Wow, was I surprised. What happened next was one of the most eye-opening and embarrassing moments in my career.

The speaker said that most of us in the room were idiots. He acknowledged that most of us also made more money in a year than he could ever dream of but that we were still idiots.

He proceeded to tell a story about the endless number of affiliates who show up at his door and how he refuses to even talk to them because they use a free Gmail account or have some ridiculous splash page as their website without any contact information. And when he goes to search for them on LinkedIn, they're nowhere to be found. Of course, he has no trouble finding them all on Facebook or Instagram posting all kinds of nonsense that the brands he represents don't want any part of.

In short, he made it clear we were unprofessional and, by default, untrustworthy.

I let out a deep sigh as I loaded my website onto my laptop. I looked around the room to see if anyone else felt the same way, and the somber mood was so thick in the air, you could cut it with a knife. I knew most of my affiliate friends sitting there didn't have a website at all, even though some of them were extremely successful. None of us were thinking like a brand or acting like someone who was building a real business.

We were all just sitting at the top of the funnel making everyone else rich and building nothing for ourselves. No wonder this guy didn't trust us. We didn't deserve it.

I'm sharing this to illustrate that in order to revolutionize your business and your life, you have to start with yourself.

Brand-building basics

Let me start out by saying you need to get your own legal advice. I am not a lawyer, and this is not legal advice. P.S., I'm not a lawyer. You probably should get your own, I don't know what I'm talking about. This is not legal advice. Get your own legal advice. Always get your own legal advice. You should hire your own lawyer.

Got it? Awesome!

Let's talk about contracts.

Almost all buyers will have their own contract. You know what most publishers don't have? Contracts. They just figure they can work off their buyer's agreement, and they rarely have a lawyer review it. This is a major problem. Like, huge, OK? Just because a buyer or network has an agreement with their insertion order (IO) doesn't mean it's good for you or for them. There are a lot of bad lawyers out there who aren't very good at their jobs. And most companies don't like spending money on lawyers, so they get the cheapest they can find.

I spend a ton of money on lawyers—writing and reviewing contracts and negotiating them—because I know how important it is. It's not important up front. It's important when there's a problem in the future.

You need to know what will happen if there's a problem down the road. (Just like a prenuptial agreement, right?) You negotiate that puppy when everybody's happy, so you're protected if someday you're both scratching each other's eyes out. Everyone is super excited to do business together in the beginning, but what happens when things turn sour?

I've read probably hundreds of agreements, and people misunderstand them. Contracts don't have to be aggressive or one sided. They just need to be well organized and provide instant clarity for both parties. If there's an issue with the contract, it's usually because the language is too simple or they've left something out.

So what should you look for in contracts and IOs?

Dispute resolution clauses. Make sure you understand fully what will happen if there's a dispute. Like if you use an ad that's not approved, what will happen? What if there's a nonpayment situation? Will you have to go to court? If so, which state will it be in? Will there be arbitration or will you have to hire a lawyer? If so, who's going to pay for it? All of that matters.

Liability and time limits. If something goes wrong, is there a limit for how much you're liable for? How long does the buyer have to tell you that there's an issue with the calls? Do they have two weeks, or can they come back six months later and ask for their million dollars back? Who's responsible for listening to the call recordings? Who decides if there's really a problem? And how much time do you have to rectify the situation.

Payment terms. All financials should be outlined explicitly with no room for misunderstanding. You will be paid every week, net seven, or every month on the 15th of the month following the calls. Make sure you know when and how you'll be paid.

Capacity. How many phone calls can you send? What's the cap? What happens if you go over? If you're not getting paid for those, it should be spelled out. What happens with duplicate calls? Can you send them to the same buyer once a month? Or never? What happens if you do? What are all the capacity terms by month, week, day, and even hour?

Campaign rules. What compliance rules will you need to follow? What are the requirements for a conversion? What are the hours and days of operation? What's the correct procedure for raising or lowering the cap?

Any agreement that doesn't have those items included is a big red flag. Usually it's not malicious or anything, just bad contract writing. That's why it's so critical for you to read all your IOs yourself and have a lawyer review them. I do that with every contract I sign just to be sure I know what to expect. Read your agreements, and if there's anything you don't understand, ask your lawyer. Also, be aware that just about everything in a contract can be negotiated, so don't be afraid to push back on a contract either.

I once had a customer who didn't have a contract reviewed, and it turned out there was a clause in that contract that said if there were any issues with the phone calls, that my customer would effectively cover the legal fees of the buyer to resolve the issue. That's huge!

Unfortunately, this person didn't find out until the buyer was sued over a phone call, and it turned into a class action suit. My customer was technically on the hook for all the legal fees, according to the contract.

Hear me: This is a totally normal clause. It's ridiculous, but I see this happen a few times a year. The affiliate agrees to cover all the legal fees for anything related to their phone calls with no liability limit. The affiliate has no say in who the attorneys are, what the settlement amount is, nothing. They just have to foot the bill. I've seen people almost go bankrupt over situations like this.

So, once more with feeling ...
Say it with me:

> I will read all my contracts.
>
> I will have a lawyer review all my contracts.

I will negotiate any unfair contracts.

I will hire a lawyer and not take Adam's word for it.

OK, good. Glad we straightened all that out. Now, just because it breaks my heart to see affiliates signing bad contracts, Ringba has spent around $10,000 in legal fees creating our own IO that any of our customers can use. I want all our people to be protected. Scan the code on the resource page in the back of the book to learn more.

Now that you've got your ducks in a row, let's talk about your business image. How are you presenting yourself to the rest of the world? That matters more than you know, and later on in the book, I'll share my most embarrassing story to show you why it's so important.

Creating a brand for yourself

It's never too early to create a brand for yourself. It doesn't matter if you're just planning to be an affiliate or you want to build your own network or broker calls. Whatever you're going to do, you need to set up a legitimate business that shows you're serious about what you do. There are too many fly-by-night operations in the affiliate world. Those don't fly at all in pay per call. You're working with established businesses and your reputation is everything.

This is 10 times more important if you're located outside the United States or Canada because you're going to be scrutinized more closely as a foreigner in this space. That's simply because the vast majority of fraud in pay per call comes from people in other countries. So, if you're not an American, you're already at a disadvantage. But you can overcome that by making a great first impression.

Branding is not a complicated process and doesn't cost a lot of money. It just takes a bit of work. If you want to see how other companies in the space do it, feel free to check out Ringba's Facebook page and LinkedIn profile. You'll see all our staff's profiles, and you can just use what we've done as a model.

Here's what you're going to need:

- A business name that shows you work in the pay per call or performance marketing space
- A .com domain name that matches your chosen business name
- A hosting account
- An email address for that domain
- A logo for your business

Next, you're going to build a website. Now, this doesn't need to be a full website, a simple one-page template is fine. But please don't put up a bunch of placeholder *lorem ipsum* text with broken links and missing information. That's not going to fly. Lots of people getting into this space throw up a template on their domain and that's it. They don't take the time to fill out all the information.

Honestly, Americans look at that and they think *Wow! That's shady. There's no way I'm doing business with them.* So, at the bare minimum, make a splash page with your logo on it and maybe a little bit of information about you and your contact information. Even if it's bare bones and super basic, just make sure it's complete.

Then you need to make a LinkedIn profile for your company. This is free and takes about five minutes. You upload your logo to it and write a short description. Connect this company to your personal LinkedIn profile page. Take the time to write a description that talks about digital marketing and pay per call advertising. Maybe even put a couple of blog posts on your website and post them to your LinkedIn profile. Your company should have a presence so that when people look for your company, they can find it and see that there's actually real people there.

On your personal LinkedIn profile, use your real name and photo. It doesn't matter if you're outside the U.S. Anything else is going to cause issues for you. Just be yourself, and you'll have a much easier time.

All your other social media personal profiles should be cleaned up. No pictures of you drinking and partying or any of that nonsense. Every now and then, it should mention something about pay per call or marketing. Use real pictures of yourself. And make sure there's nothing political or religiously motivated on your wall. You want a nice, clean public profile that shows what a nice person you are.

On Facebook, you should also set up a business page with your logo, a cover photo, and a few posts about your business. Get some friends and family to like it to show people it's legit. Fake names and photos will turn potential networks and business partners away. Just be yourself. The reputation you're building now will be very valuable in the future.

Here's another quick tip for anyone outside the U.S. You need to make your emails sound professional. That means you should have an email signature that includes your company name, your office phone number, and maybe your social media URLs. You might also add a business photo of yourself to make it all feel more friendly.

If you don't have an American phone number, register for Ringba, generate one, and forward it wherever you want the calls to go. You should have a North American phone number that people can actually call that connects to your actual phone—or at least a voicemail. You can set up that voicemail in Ringba. It's super easy to do.

With all that set up, you have a professional online presence. And when you start to communicate with people in the space, they're going to Google your company name, they'll look at your website, they will see you have social media profiles, and they'll see that you've taken the time to at least set yourself up as a legitimate business. After that, they're going to be much more likely to work with you.

I know this is very basic information, but a lot of people don't do it. I don't want anything to stand in the way of your success.

Set up an LLC or corporation

To officially establish your company, you'll need to set up an LLC or corporation. There are lots of reasons to do this, including limiting your personal liability and potentially lowering your tax liability. Let me start by saying, again, I am not a lawyer. So when it comes time to set up your business you should talk to either an attorney or an accountant to get the best advice on how to go about it.

Setting up a legal business is not too complicated and it's not expensive. Mostly, it's filling out a lot of forms. You can check your state's website to find out how to register. Or there are lots of companies online like LegalZoom or Matchstick Legal that can help you.

You don't have to live in the same state you register your business in. Many business owners register in Wyoming, Nevada, or Delaware because they have fewer regulations and no state income tax for corporations. Again, do your research, and get an attorney to advise you if you have questions. The state websites will walk you through the process if you decide to do it yourself.

You're going to need something called a registered agent, which is just someone who can collect and sign for any official mail that may be sent to your corporation. Every state has companies that are allowed to act as registered agents, and they usually have a mail delivery service available, too. They'll even help you set up your paperwork for a small fee. You can also be your own registered agent if you're setting up your company where you live.

Once you have your company established, you're going to go to the IRS website and get an EIN. That's an Employer Identification Number. You'll need that to open a bank account, to run payroll, file your taxes—everything financial involving your business will need the EIN. All you have to do is fill out a form online and they'll assign you a number. It's free and simple to do.

Next you're going to set up a business checking account with a major national bank. Don't use a small regional credit union, because many of them can't accept wire transfers from international banks. It can become a real hassle if you want to get into international pay per call. You're going to get a corporate debit card with your account. This is what you'll use to buy anything associated with your business. Always keep your company finances and your personal finances separate. Only use your business debit card for business purposes.

Eventually, you're going to have legal and financial questions that you really need a professional to answer. Start looking around for an attorney and an accountant so that you have them available when you need them. They don't have to be expensive, but when you need one, you really need one.

So that's it! You've got a great looking brand, an official business, and you know which vertical you want to get started in. You've sifted through offers and found a couple that look promising. Now it's time to build your first campaign.

Let's get started ...

CHAPTER SIX

BUILDING YOUR FIRST CAMPAIGN STEP BY STEP

My first campaign? Total disaster! Like, an absolute shit show. I already told you what happened with the merchant account buyer—I totally flooded his call center with unqualified calls and burned down their revenue for an entire day. I thought I knew everything about marketing, but I was thinking like a traditional affiliate. I was thinking digital, where unlimited clicks were the best possible outcome.

You'd think that one episode was enough, but no … I'm a glutton for punishment. I did it again with another direct buyer before I learned my lesson. (And no, I didn't have call tracking for that one either. It still stings a little.)

You're going to hear me say this over and over: You're dealing with humans, and humans have limited capacity to handle traffic. My biggest mistake was not understanding capacity—like at all. And neither did the buyer. They kind of gave us a monthly cap, but I didn't know to ask for a concurrency and hourly caps. So a huge percentage of our calls just dropped. I didn't know how many, didn't have any oversight, and didn't really know what was going on. We were just flying blind in a blizzard and hoping everything worked out. It was a terrible feeling!

Fortunately, that campaign was profitable. But all the mistakes we made cost us about half our potential profit because we didn't have multiple buyers and we had no idea how many calls were dropping. I don't want that to happen to you, OK? The only way I can live with my god-awful mistakes is knowing they will benefit someone else. So let's get started building your first campaign. And smile, because it's going to be so much better than mine was!

Be a beginner

OK, you've picked a vertical and approached some networks. They've accepted you because you showed them you're a professional. I recommend having a few networks right from the start, if possible, so that when you build a campaign that works, you can start load balancing right away. Remember, just because one network pays more money per call doesn't mean you actually *make* more money per call. You can't do the math to figure out who has the best buyers until you can load balance them against each other.

Now it's time to get down to building.

Once you've talked to your account manager, and they have given you three to five offers, you're going to focus on a specific traffic type. Before you choose, though, you're going to ask if the network has any landing pages already created that you can use, say, maybe AdWords call-only landers. That way you don't have to create anything, and you can just focus on generating calls.

If you need to make your own landing pages, hit up Ringba. We have tons of them, and we give them away for free, both long-form advertorial style for social media and click-to-call pages for search marketing. Use them for whatever you want, modify them or not—we don't care.

Networks aren't really in the business of supplying their publishers with materials, but we happily give them out because we want everyone to be successful. The more people out there using great materials to drive calls, the happier the buyers are, and the more money everyone makes. Win-win-win. Just how we like it.

If you do decide to make your own landers from scratch, make sure you get feedback from your pay per call groups. There are some amazing experts who will give you honest advice on how to improve whatever you've got. Run your ad copy by other marketers, too. If you're worried about people copying you, don't be. They can, and if they want to, they will.

It's not top secret information. There's nothing you can really do about it, so you might as well get some free advice.

Whether you show your ads or not, people who know what they're doing will always be able to copy it. But the real pros know what they're doing and aren't really interested in copying anyway. You have nothing to lose by showing your first couple of ads to people. Get feedback from professional marketers. I can't stress this enough: Most people who go into our groups and say things like, "I can't make pay per call work. What do I do?" No one responds to them because they're not actually trying.

But if you take your ads and expose them to the community and say, "Hey, folks! How can I improve this?" You'll inspire the new people, and the veterans will help you. If you post it in the Pay Per Callers forum, and I happen to see it, I will personally give you feedback. So will our team, and a bunch of networks, and so many other people. You just have to be open about it. Then you're going to change that creative based on their feedback. At the end of the day, you're not actually exposing what you're doing because those aren't the ads you eventually run anyway.

People can be incredibly generous if you just show them the material and give them the information that actually allows them to give you feedback. There's nothing I love more than helping new people succeed in this business. That's why we set up the Pay Per Callers forum in the first place. We're all happy to help inside the forum, but very few people are happy to show what they're working on so that we can actually give advice. Just don't be so uptight about protecting your work, OK? You're not doing yourself any favors.

Establishing a testing budget

OK! So how do you decide what your testing budget should be? First of all, realize you're probably going to lose money. Get comfortable with that. Just make it a game and assume you're going to lose most of the time. That makes the whole process much less intimidating.

There are a few ways that you can creatively generate call traffic without any money. Find places to post where people congregate, or you can manually hit people up through direct messages. It is possible, but it's not usually scalable. If you have to start there, that's OK. But save your money so you can start buying ads as soon as possible.

If you're going to start with Google AdWords or Facebook ads, you're going to need about $500. That's the absolute bare minimum if you have some type of background in these spaces and kind of know what you're doing. If you don't know anything about pay per call, and you only have $50 to try and figure it out, you're going to have a really hard time doing that, so save your money and then try harder later. Realistically, with $500, you probably can test one traffic source. It probably should be Google AdWords, and you're going to move really, really slowly.

If you're going to buy media and get really good at it, you have to realize that your time is more valuable than the money. For instance, if I were going to test a campaign on Google or Facebook, my minimum budget would be around $1,000 for a single campaign. I would probably create 100 different ads, and I would try to burn that money as fast as possible to see what users are reacting to.

Then I would throw another $1,000 at a completely different group of ad sets, and then another $1,000. For me to really learn a campaign, I'm probably going to light $3,000 on fire. Then I'm going to spend $5,000 optimizing, and *maybe* I'm profitable at that point. Realistically, I'm going to throw $10,000 at a campaign—and I'm going to do it as fast as I can.

That doesn't mean you can't do it for less. You can. But I'm going to do it really fast. That's the kind of money you need to throw around if you want to build out, develop, and run a new campaign at volume in just a few days, especially on social media. If I'm running Facebook ads, I'm going to spend $10,000 on my first day because I want to find the win as soon as possible.

You can absolutely be patient and spend less money and build your campaign slowly, but if you want to move fast with these campaigns, you need a bankroll to do it. That's why it's really important to save your money.

Now, keep in mind that when I say "lighting money on fire," that doesn't mean you're not going to get any revenue back on it, and it doesn't mean you're not going to learn something that changes your whole business. That's why I'm so willing to spend the money on media to figure it out, because I'll learn something that will change everything. I'll probably make some of that money back, and so will you, but assume that you're just going to lose all of it because then you can make financial decisions you can live with.

DO NOT take your life savings and blow it on buying traffic so that tomorrow you have no money and you're screwed! (Read that again, please, just to make sure it sinks in.)

Set your risk tolerance in accordance with your capital. If you're an affiliate coming over to pay per call from another vertical, you're going to have more capital, you're going to understand how this works, and you're going to be OK losing some money at first.

But if you're brand new to this and you need to figure it out, you probably want to start slower and not move into the more complicated mediums like out-of-home, print, radio and TV yet. You're just going to lose a bunch of money without learning from it simply because you're not sophisticated enough to do so at this point.

The first time I ran a TV ad, I ended up losing 100 percent of my investments. It was just an abysmal, absolute failure. The commercial was hilarious, and we had a lot of fun making it. And later we went on to build a successful TV commercial, but the first one we put out there was just a failure. You should expect that. Learn to laugh at yourself. You're like a little kid figuring out how to ride a bike. It's hard at first, but you'll soon get the hang of it. So don't take everything so damned seriously, OK?

How do you determine how much money you're actually going to need? You need to understand what your target cost per caller is. If your campaign pays out $25, then your target cost per caller should probably be $15. That would give you a healthy margin and a healthy amount of money to buy clicks somewhere in the middle.

Multiply whatever payout you're getting by three to get an online potential spend per source. If you start an AdWords campaign, and you're being paid $10, and you buy $30 worth of clicks to your lander, and you get zero phone calls—you know something's wrong.

Around 300% of your payout should be the upper bounds of the limit for you to learn something.

Now, if you spend $30 and you get one phone call, that's great! It means you need to start optimizing. But if you spend $30 and get zero, something is wrong or you're approaching it the entirely wrong way. This is just a simple rule of thumb. If your payout is $100 and you spend $300 and you get nothing, you're really in the wrong spot.

Here's the thing, though, conversions can come in waves. If your payout is $10, and you spend $10 and get zero, you might've gotten three phone calls if you had just spent $14. That's kind of how advertising works. It's funny like that. So, when you're working with micro budgets, your results aren't going to be typical. But the more money you spend, the more likely that your results will be typical on a campaign because you have more data to gather statistical significance.

Choosing traffic sources

At some point you need to decide what traffic sources you're going to work with. The simplest way to do that is to research other people's campaigns. Do a little ethical spy work. Google around, search Facebook for companies in the industries you want to work in and click the "info" and "ads" tabs on their Facebook pages. It's really easy to spy on anyone on Facebook and Google. You just go search around for what you want to promote, and the answers on how to do it are given to you. It's not rocket science. You can just go look.

Spying does not mean copying. Have some dignity and ethically model what's working. Don't steal it. That's just shitty behavior.

Next, research your keyword cost-per-click so that you understand how much it's going to cost you to advertise. If it's $80 a click but your payout's $30, you should not run that campaign. Look for another option. But, if your payout is $10 and it seems like clicks are going for $0.50 to $2, you can figure out how to make that work. It's all just simple math, and if you do your research up front, it's not that complicated.

Start testing

You've got your landers and ads, now it's time to run some tests. Start by testing small to see if you can generate calls. If my budget's restricted, this is what I'm going to do. I'm going to test, test, test, test, test, test, test until I generate a phone call. Maybe I'd take all three to five offers at four different networks, and I throw $15 per campaign, per network to see if I can get some calls generated.

Now, I don't know if you've figured this out about me yet, but I'm a freak. I absolutely love working in this space. It's fascinating to me. The most amusing thing I could possibly do is create advertisements. So, I'm going to create hundreds or thousands of ad combinations. You'd better believe if I'm running a social campaign, then I'm testing thousands of combinations to figure out what works.

The difference between a successful affiliate and someone who fails is their willingness to test a ludicrous number of ads and be willing to lose on most of them. Because when you hit, you hit *big*—like life changing big in a matter of seconds.

A lot of affiliates are like, "Well, I created two ads and I got eight keywords, and I used someone else's landing page. But I'm not really making any money. What am I doing wrong?" First and foremost, they're lazy. They should've tested 300 advertisements, 400 keywords, and five different landers. You need to optimize everything, and it's all about testing, testing, testing, testing, testing. (Did I mention testing? OK, good.)

Your minimum budget per campaign or keyword is roughly 300% of your payout, you just have to accept that. If you can't get to 300% of the payout, you're probably not going to succeed. Now, divide your budget among multiple offers and traffic sources like we talked about to reduce your risk. If you have a $1,000 budget, run that in twenty $50 blocks so that you have a lot of chances to get some traction and then reinvest whatever you make into more testing.

When you're new, part of your job is listening to 100 percent of your calls without exception. Just have them playing in the background while you're working. You're going to catch some crazy shit you never thought about when you were building the campaign. Someone might call in asking how to apply for a job, but it's a medical device campaign. That should make you stop and think, *Crap! I need to rethink my negative keywords.*

You're listening to better understand consumer intent so that you can modify your marketing to get the call quality up. If you send 40 people calling about employment to a call center for a campaign that's about medical devices, the call center's going to bitch, the network's going to shut you down, and you're not going to make any money. Just listen to your calls. It's amazing what you'll learn.

Expand relationships and load balance calls

Once you have active campaigns that are stable, and you're generating maybe 30 to 50 calls a day, the next step is investing time into expanding relationships and load balancing your calls. Do it as soon as possible. Work with every single buyer and every single network humanly possible to find out who's brokering what to whom, cut people out of the equation and add more buyers into your own routing plan. The goal is to see which buyers make you the most money and to expand your capacity.

You find the insurance office—like an Allstate representative who has three salespeople and wants phone calls—and you work with them. They're licensed to buy the calls and sell the insurance, and if they just had more phone calls, they could hire more reps.

It's more work to get your own buyers, but once you get them, your margins are much higher, and that means you can outbid everybody for the traffic. And when you can do that, no one can compete with you. That should be your end goal.

What you'll learn very quickly is a lot of offers are brokered in the pay per call space, and you can navigate your way through to the direct buyer, form an exclusive relationship, and bump your payouts up. Direct buyers would rather work with direct affiliates than networks 100 percent of the time, so you have a competitive advantage.

If you can get direct to the buyer, and you have a history in that campaign of delivering quality calls, as well as recording examples and quality-assurance checklists you can show them, they're going to be happy to work with you.

You can start with just a few while you figure everything out. Maybe you're only selling them a couple of calls a day. That's fine. You can send the overflow traffic to the other networks. You're slowly rising by building a network of micro-buyers through a repeatable process. And once you get the process down, you can hire business development people to do it for you at scale.

Create Your Own Brands and Assets

As if you don't have enough to do, you'll want to start creating individual websites for each vertical you're in. This isn't the same as building a brand for yourself. This is your play to own the whole freakin' vertical by building a powerful brand that everyone will trust and use as the authority in the industry.

First, buy a short, memorable .com domain name with no hyphens or numbers, and no funny puns. It has to be something obvious that people will remember like *calladentist.com* or *insurancehelp.com*. We found Ringba.com for $9, and now we have a fantastic brand built around it. Your goal is to build high-quality websites and social platforms.

- Hire a writer and create some great content for each vertical.
- Build social media profiles and post content to build a following.
- Invest in SEO link building for your site.
- Drive traffic to your landing pages on the site.
- Buy expired domains and use the Wayback Machine to recreate the content from those domains.
- Create a whole network of websites in each vertical so you can dominate the search results and mind-share in that space.
- Use Flippa to acquire sites that need some polish, fold them into your business.

It's going to take a little while, but it will be worth the effort. Once you've established your own brand in a vertical, then it's time to start expanding your advertising into TV, radio, print, and out-of-home media. With those methods, you'll be saturating geographic areas and building something that has enterprise value.

Now you're vertically integrated. You own the whole path—the traffic, the buyer network, the recognizable brands—and when that happens you can get acquired for serious money. Maybe you go live on a yacht or travel the world for the rest of your life. Me, I'm going to start all over again in another industry because this is just what I love to do.

We've discussed quality assurance a little bit, but next I want to share with you exactly how to use it and how to avoid falling for fraud.

Ready?

OK, get a snack first. Then let's go ...

CHAPTER SEVEN

COMPLIANCE AND QUALITY ASSURANCE

I couldn't believe what I found when I started actually listening to calls. It took me a while to get my own call tracking technology, and when I did I was so excited to hear the actual conversations that were happening. At first, I was only listening to the longer calls because my goal was to learn more about consumer intent and the questions that were asked over and over. I wanted to build better ads, and that was about it.

But I couldn't understand why there were so many shorter-duration calls. Those short ones never hit the conversion duration, so we weren't getting paid for them. So just for fun, one day I sorted the calls and started diving into the ones that came in under the conversion duration. And that's when my head exploded.

I heard the standard greeting and then the call center agent said, "Can you hear me? Hello? I can't hear you very well. You're breaking up. Can I call you back? I'll use this same number. Is that OK?" Of course, the caller agreed. They hung up and the agent called the guy back.

Whether they made a sale or not, guess who didn't get paid? Me! I had accidentally stumbled across the most common form of buyer fraud in pay per call—callback fraud. The agents weren't obvious about it. They weren't calling back every inbound call. But roughly 20% of the time, I was losing a commission.

Now, I had no protections in place in our agreement for this, and I had no real recourse for recovering the money. Obviously, I stopped working with that buyer, and I did dispute the issue. They apologized and covered a little bit of my losses. But mostly they made up some lame excuse and went on with their day. These days I always have provisions in my agreements to cover this kind of problem.

Publishers never want to think about compliance or quality assurance. They believe it's a cost center and a waste of time. But in reality, participating in the compliance process is how you preserve access to buyers and offers. It's a way to preserve your reputation in the space and build a real long-term sustainable business. Too many affiliates are fly by night. They think if they just make their advertising really aggressive, they can make some money and everything's all good.

But what actually happens is they lose access to campaigns. They don't get paid. They get burned. They're left holding the bag, and then they wash out of the industry. The difference between following a compliance process and not following a compliance process is not getting paid and being put on a block list so no networks or buyers will let you in. Don't let that happen to you. Compliance is not that complicated, OK?

When you're working with any buyer in any industry, there are rules you have to play by. Usually, these are rules set up by government regulatory agencies, and the buyer can get into big trouble and be sued for millions of dollars if you screw up and don't play by the rules. That's why they are so picky about who they will allow to send them calls. It's a serious thing.

So what do you have to do to play by the rules and stay in compliance? Rule number one, and really the rule that matters most, is to be transparent. Tell them everything you're doing. Many buyers will require you to show them your ads. Like, you cannot run the ads without prior approval. So just show them the freakin' ads. It's not a big deal. Some publishers will try to be sneaky and get one ad approved but run a different one. It's not worth the risk.

A lot of publishers don't like the ad approval requirement because they're afraid of people stealing their ads. If you feel this way, get over it right now. All ads online are public. Millions of people see them, and anyone can go to the Facebook or Google ad library and find your ads. If they want to rip you off, they can. And there's nothing you can do about it. So the best thing you can do is just get over that fear and play with the pros. The pros don't worry about stuff like that because they're too busy making money.

The worst thing that might happen is a buyer says, "Yeah, you can't use that ad." But that's fine because they will tell you exactly how to fix it to make it compliant. You're basically getting free legal advice and keeping both you and your buyer out of a legal dispute.

Another huge no-no is sending third-party traffic. In other words, you're brokering calls from other affiliates to the buyer. That's a problem because the buyer has no idea what the original affiliate was doing. Were they legit and using an approved ad? It's hard to say. If anything illegal was going on, the buyer would be in trouble. So read your contracts carefully and make sure you fully understand all the rules of a particular offer. If you aren't sure whether you are allowed to sell third-party or affiliate-generated calls to a buyer, ask! It's better to be safe than sorry when it comes time to get your payment.

The buyer isn't trying to be mean, they're just trying to keep from being sued by the government. And trust me, you want that, too. The rules are there to protect everyone.

Quality assurance

When you're listening for quality assurance, you're checking for quality of both the traffic you're sending *and* the buyer's agents. The only way to do that is to listen to your calls, and you must have your own tracking software to do that. There's no way around it. And, really, don't you want to control your own destiny? Yeah, I thought so.

It's so important because it keeps everyone honest. Even if you don't think anything malicious is happening, you should still run quality assurance checks just to be sure. After all, if your conversion rates are going down, how can you possibly tell whether it's your targeting, your ads, your creative, or just maybe something's going wrong on the buyer's side of things? You can't unless you're doing QA regularly.

In large-scale operations—we're talking thousands of calls across 10, 50, or 100 campaigns—you're going to find issues. You just are. So you need to understand what those issues are and how to deal with them. I always tell my students to assume the best of people. Most of the time there's nothing malicious going on, just human error.

You're working with thousands of humans who make their living answering the phone. Not everything is going to work smoothly. So work with your partners to resolve issues without a lot of stress or anger. Your blood pressure will thank you. If you communicate clearly and resolve the issues with minimal drama, you're going to forge strong relationships, and that will grow your business.

It's not just about checking for fraud, though. When you do this process right, your call value goes up, your conversion rates go up, you widen your profit margins, and buyers may even be willing to pay you more money.

Another great benefit is that you're going to impress future partners and networks. Not all publishers have a quality assurance process—most don't even do their own call tracking—and that makes them less desirable as partners. But when you can demonstrate that you take this business seriously, they're going to want to work with you. And if you're working with direct buyers, rather than through a network, they'll be willing to pay you more per call just because you're approaching the industry in a professional manner.

Publishers get the most upset when they don't get paid for certain calls. Whatever you do, don't go bashing the buyer or screaming about them on social media. Resolve the problem calmly and ethically. Your behavior directly affects your reputation and your future in this business.

Before you sign any agreement with a network or direct buyer, make sure you know exactly what they're paying you for and what conditions will result in no payout. If they don't point it out in the agreement, ask them. That way you'll understand exactly what went wrong when you don't get an expected commission.

This part of the business is a fair amount of work, and you're going to want to hire someone to take care of it once you start to scale. Fortunately, quality assurance experts are easy to hire or outsource, and they're not expensive. But before you outsource, it's critical that you understand how to do the process correctly. So, let's take a look at some common problems that crop up.

Duration billing and callback fraud

Callback fraud happens when a call center agent detects qualified prospects very quickly and comes up with a reason to hang up and call the prospect back. Maybe an agent says something like, "Oh, I can't hear you, sir. Let me call you back real quick." Then they dial the person back using their caller id, which means the call is no longer going through the call-tracking platform. Since the call doesn't reach the required duration to trigger a payout, the publisher doesn't get paid.

So the buyer trains their agents to steal random phone calls from the publishers because they know the vast majority of people involved in pay per call do a really poor job with quality assurance. The buyer will likely never be caught.

We see it all the time. When a publisher customer calls us wondering why their conversion rates have dropped, I always tell them to listen in on some call recordings. And guess what we find? Callback fraud. It's incredibly common.

This is why I tell publishers that quality assurance is a profit center. You have to listen to the actual calls to find out what's really going on when your conversions drop. Otherwise, you might screw up a perfectly good ad campaign because you think something's wrong on your end, when, in fact, someone isn't playing fair.

If you'd like to hear some actual examples of what real-time callback fraud sounds like, just follow this QR code to some actual examples.

This happened to a client of ours recently on a dental ad. Someone called a dental company and the call center agent said, "Is there something you need right away?" When the caller said he needed braces, that agent knew this was a high-value call. There was some distortion on the line, and the agent told the man to hang up, and he would call back.

This was a 35-second phone call, and the minimum duration for a conversion was 90 seconds. So in a period of 35 seconds, the call center agent was able to identify the caller as a high-value target, hang up the phone, call the guy back, and get the phone call for free.

The call center wasn't on the hook for paying for that phone call but yet reaped all the value out of that phone call. We later found out that this buyer was engaging in this activity all the time and had been for a long while, but our client wasn't listening to the calls, so they had no idea it was happening.

It's sad, but if buyers don't know you have a QA process in place, they're probably going to look for ways to get around paying for calls.

We've noticed that, in general, the smaller the buyer and the fewer the number of people in a call center or phone room, the higher the chance that they will engage in callback fraud.

We define a small call center as having between 10 and 30 agents. With such a small capacity to take calls, the buyer only needs to rely on a couple of publishers to drive calls. So it's highly profitable to just train their agents to do call backs and not pay the publishers. If they leave, so what? There are plenty more pay per call publishers out there. They can just burn these people over and over.

The bigger buyers usually don't engage in this kind of activity because they have significantly more money invested into their infrastructure, their employees, and growing their business. For them, the risk of losing quality partners over a little bit of extra profit is not a good business plan. It's just not in their best interest to churn through partners because it takes time to find new ones, and time is money.

How do you catch callback fraud?

You're going to have to listen to call recordings. I find it fascinating, but most people don't want to invest their time going over insurance and pest control sales calls. But it's so important. Not only so you can catch anything sketchy that's going on but also because you can learn so much about the industry. That knowledge can help you create better ads and build higher-value campaigns. If most people won't do it, then all the better for you.

You're going to have thousands of calls going through your tracking system, and you certainly don't want to listen to all of them. The fastest way to do your QA to catch callback fraud is simply sort your calls by duration. You can usually skip the really short 5- to 8-second calls. But anything longer, you want to check. If a certain campaign converts at 90 seconds, then you want to listen in on calls between 9 and 89 seconds. Start with the longest and move to the shortest on your list, and listen to all the calls for one buyer at the same time. That way you can tell if it's a pattern for that company.

Keep good notes, because when you contact the buyer or network to demand payment, you're going to need proof. If you see this kind of behavior more than once with a particular buyer or network, you also want to request that penalties apply. It's not a fun conversation to have, but they need to know you're serious and that you expect them to do their jobs.

What if the publisher is the one committing fraud?

Yeah, don't be that guy, OK? Just don't. It's easy enough for buyers and networks to figure it out on their end. Smart, savvy publishers will have more than enough call traffic and will be making plenty of money. There's just no reason to be making calls yourself or sending in mystery shoppers making fake calls. That's just wasting everyone's time, and it will ruin your reputation if you want to sign up for future campaigns.

There are a few ways publishers commit fraud, and I want you to know about them so you don't get caught doing something you shouldn't.

Incentivized calls are when the caller is promised something of value in return for the call. Something free doesn't count. But if you're running ads promising points, merchandise, or even cash as an incentive to call, that's fraud. Most contracts you'll sign will specifically say no incentivized calls.

Mystery shoppers are just what they sound like. A publisher pays someone to make a bunch of calls pretending to be an interested prospect. They hang on just long enough for the call to convert and then hang up. That's not cool. Don't do it.

Sometimes publishers will even try to fake data inside of an IVR and manipulate the answers to beat the filtering. This will be found out sooner or later, and it's really not worth the effort. You're far better off just staying in compliance. As I've mentioned over and over, there's plenty of money for everyone.

Noncompliant warm transfer scripting.

OK, this one's fun. This is when publishers use noncompliant warm transfer scripts before they transfer the phone call to the end buyer. Whoa, that was a mouthful … let me back up a bit.

The buyer only wants the most highly qualified callers, right? So, sometimes a publisher will warm up and qualify those callers by sending them to a separate call center first. The caller answers a few questions, and then the agent transfers the call to the actual buyer's sales team. Since those callers are now warm leads, we call it a "warm transfer."

But there are very specific things you can and can't do when transferring a caller. Every campaign is different, but in a nutshell, you don't want to be transferring callers who don't fit the customer profile. Maybe you're running an insurance campaign, and it requires people over the age of 30. That's easy, you can qualify them with the ad. But maybe the callers also need to currently have auto insurance, have no accidents in the last two years, and no DUIs. It's trickier to get full qualification with just an advertisement, so you direct them to a warm call center where an actual person does the qualification and then hands the right prospects off to the buyer.

If callers don't qualify for the product or service, you don't transfer them. Simple.

Publishers who transfer unqualified callers knowingly are not only committing fraud, they're also just plain shortsighted. They do all the work to drive the calls and set up screeners, then they just get lazy and try to keep people on the line long enough and then send them to the buyer. Sooner or later they get caught. They lose the payments on those calls, they lose their sunk advertising costs, and lose their reputation with other partners.

With just a little bit of extra attention, a couple of questions on the screening script, they can make those warm transfers compliant and keep everyone happy. It's easy enough to make money in this business without defrauding people. So, don't be lazy, OK?

Poor sales agent performance

Sometimes no one is doing anything wrong, but the sales team just sucks. This isn't really fraud, but it can still be a big problem for publishers who don't do QA. You can send all the traffic in the world to a call center, but if they can't close the sale, everybody loses. You lose a commission, the buyer loses a sale, and the customer loses out on a potentially great product or service.

If you're working with a network, you'll want to point out the poor sales performance to them. They may be able to go in and help the sales team improve. That's a great opportunity for you to show your value to the network and the buyer.

If you're working directly with a buyer, and you know how to train salespeople, then maybe you offer to go in yourself and help them learn to convert at a higher rate. Sometimes, the buyer doesn't even realize their sales agents need help. You're doing them a favor by sending over a few calls to show them how much they need real sales training.

Either way, you don't want to be spending a lot of time and money driving calls if most of them don't convert into customers. It's just demoralizing. So, make sure you're prioritizing the high-converting call centers.

Agent cherry picking

Sometimes sales agents make their own decisions as to which calls are going to convert to a sale and which ones aren't. Maybe in their experience, women spend more than men, or older callers take twice as long to make a decision, or certain ethnicities convert more often. The agent knows more calls are coming, so they don't want to waste time with the people they don't think will convert.

The agent is essentially cherry picking the best calls and burning the rest—which means if it's under the duration limit, you don't get paid for driving a legitimate caller. Plus, the buyer loses out on a potential sale, all because the agent wants to talk to the "winners."

You'll know this is happening because a particular buyer's percentages are out of whack. Like, their agents hang up on 30% more callers than everyone else. In this situation, you need to let your network or buyer know what's going on. Make sure you have proof, including the agent's name and what pattern seems to be happening.

This situation is another reason you want to route callers to several buyers at once. That way if one is misbehaving, you can just adjust the flow so your calls avoid the troublemaker. Even so, you should still be paid 100 percent for cherry-picked calls. Most networks will state this explicitly in their contracts and will penalize the cherry-picking company to keep them from continuing.

Have I convinced you that you need to listen to your calls? Great! To make life easy for you, I have a whole analysis system and QA Grid waiting for you inside the Pay Per Callers forum.

It's time to start running ads. Finally! This is my favorite part of this whole business. My ad philosophy is to go where no one else is willing to go and do the work others are too lazy to do. There's an incredible amount of opportunity out there in places you'd never think to look. The next several chapters are devoted to the traffic sources I love best.

But before we get to those, let's spend a little time understanding the psychology of advertising. It's going to help you really *get* your callers and what makes them pick up the phone. And it's going to make your ads a million times better.

You in? Let's go ...

CHAPTER EIGHT

THE PSYCHOLOGY OF ADVERTISING

My favorite example of an ad that used psychology brilliantly was a clip my friend Matt filmed and used for all sorts of ads. He literally followed a short, fat man with a giant bald spot into a Costco. That was the whole clip, but he used that as the introduction for all sorts of ads. They would open with …

"Hey! Just followed my accountant into Costco …"

"Hey! I just followed this guy into Costco because I saw him getting out of his car …"

"Hey! I saw this guy drop off his kid, so I followed him into Costco …"

The video zoomed in and out, parts of it were blurry, and Matt changed his voice and how fast he spoke. All of this was done intentionally to drive curiosity. In this age of 10 billion advertisements hitting us all at once, we have maybe two seconds to capture someone's attention. And the best way to do that is to hook into our psychological need to satisfy curiosity. People watched those ads and followed the call to action because they just had to find out what this random guy in Costco did! We can't help ourselves.

Great advertisers use images and videos of things that may have nothing to do with what they're selling, specifically because they drive that curiosity. And people have to scratch that itch. So if you can make someone curious and then get the fundamentals right, like the copy and call to action, you've got a winner on your hands.

But curiosity isn't the most powerful psychological tool you can use for advertising. Purchasing decisions are almost always driven by emotions. People don't typically pick up the phone or make a purchase unless they *feel* a reason to do so. Even very logical people are still driven primarily by emotion. The logic backs up the decision and makes the brain feel confident it made the right choice. But feelings rule.

As marketers, it's our goal to trigger an emotion in somebody to get them to take action, which means selling only the features of a product or service is a recipe for failure. Features don't get people excited or help them connect with your brand. And they definitely don't move people to action, which is the whole point of advertising.

I just want to preface everything in this chapter by saying that marketing, in general, does not work on a one-on-one basis. The advanced techniques that I'm about to show you won't work on individuals. But that's not our goal. Our goal is to create advertising campaigns that make larger groups of people feel a certain way based on emotions that they've experienced in the past. What we want to do is tailor our advertising so that it applies to people in our audience in the most general way possible.

All human experience can be generally categorized as positive or negative. For instance, we all basically feel the same way about going to the DMV. It's not fun. People are grouchy. It's hot, and you're waiting in line behind a hundred other people. And there's a very real chance you'll finally get up to the window and they'll turn you away because you didn't have the correct paperwork. It's not a pleasant experience. And so, if an advertiser wants to reference the emotions of frustration and annoyance, all they have to do is trigger a memory of the DMV in our minds and we will feel those emotions. We can't help it.

People are bombarded with all sorts of imagery, music, video, and text every day. Smart marketers know how to utilize those tools to create a desired outcome. Advertising isn't rocket science when you understand the psychology behind it and the process for building a successful campaign.

As marketers, we're effectively memory miners. All we have to do is dig up the correct memories and emotions, and bring them to the forefront of our audience's consciousness so that they feel those things again.

The Psychology of Advertising   97

I'm going to make this as simple as I can, and you might skim through this section and think *yeah, yeah, I got it*. But I encourage you to read it over a few times so you *really* get it. Because if you can master advertising psychology, you will be light-years ahead of any competition.

Using Plutchik's Wheel of Emotion

A psychologist named Robert Plutchik came up with the Wheel of Emotion to help organize the complexity of the human experience. His idea is that there are only a few base emotions, and when we combine certain emotions together, we experience different emotions. It's kind of like there are only three base colors—red, blue, and yellow. But when we combine them together in different amounts, we can create hundreds of different colors and shades.

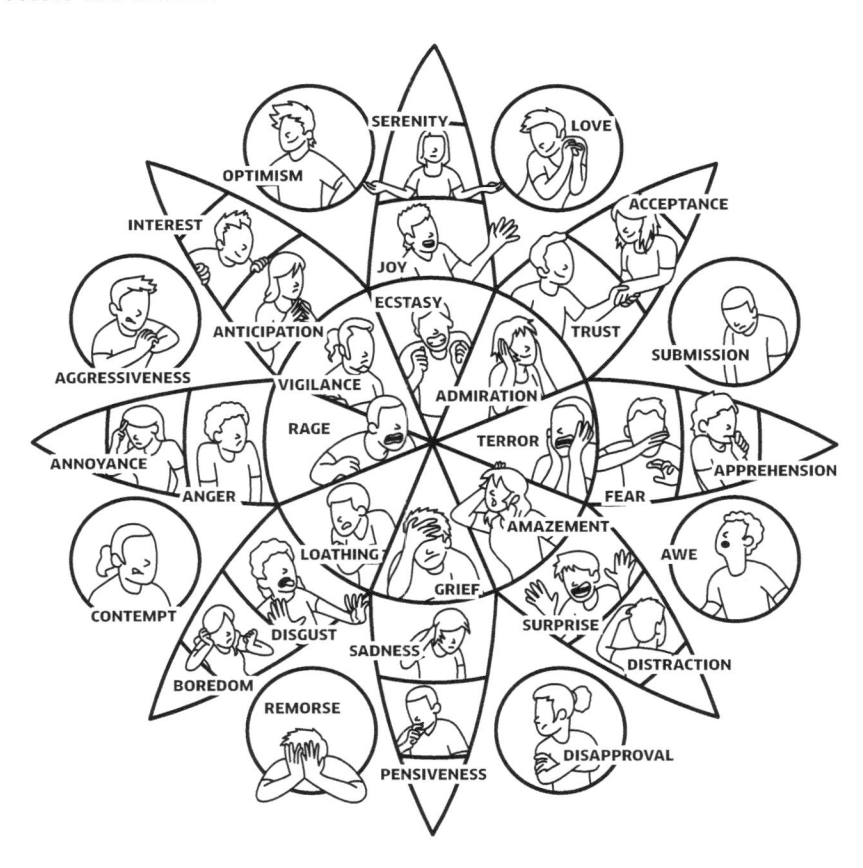

98   The Pay Per Call Revolution

For instance, at the root of annoyance, we have rage, and at the root of acceptance, we have admiration. At the root of serenity and joy, we have ecstasy. The root emotions are the most primal and powerful ones we have. Some are negative and some are positive, but for advertising purposes some studies show negative emotions are much more likely to evoke a response and get someone to take action, so you'll want to focus on those.

Plutchik's wheel is a little too simplistic for actually writing ad copy, so I put together something far more robust. It doesn't exactly match Plutchik's wheel, but it's close. In this wheel, there are over 100 different emotions that people feel and then root emotions that tie them all together. You can just use this wheel whenever you craft advertisements.

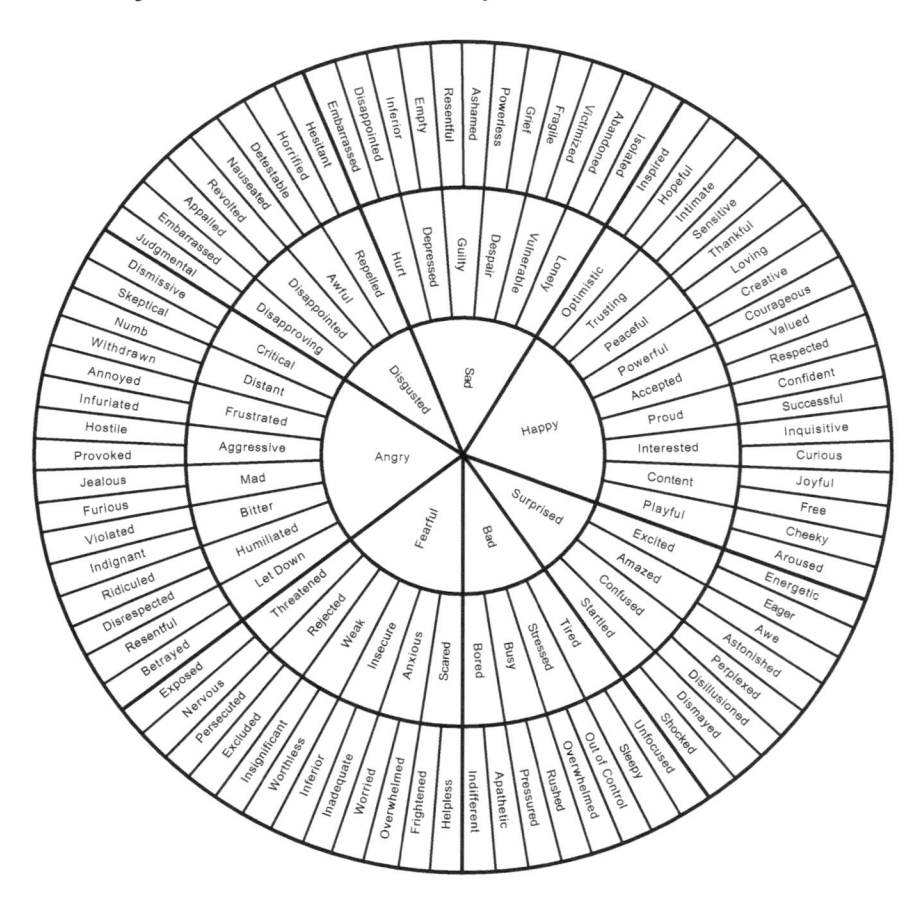

Understanding emotional cause and effect

Every single emotion that a human feels has a cause and effect. Here's how the process works. First, there's some sort of stimulus. Then our subconscious evaluates that stimulus event and decides what it was, whether it's dangerous, how we feel about it, and whether we need to do anything about it. This all happens in milliseconds without us being consciously aware of it happening. If you're afraid of snakes, and you see one off in the bushes, you're the hell outta there before you even have time to think. That's your subconscious reacting to the base emotion of fear and danger or threat.

After we feel the emotion—fear, joy, apprehension, or whatever—then we take action or modify our behavior based on the outcome we desire. If the stimulus is a lion standing 20 feet away, our initial response might be surprise. And we could take a few different actions based on our desired outcomes. If the response was fear, we might run in an effort to escape being mauled. If the response was delight, we might snap a dozen photographs in an effort to record the moment of a lifetime. Or we might think *dinner!* And attack the lion head on.

OK, not many of us are likely to meet a lion face to face, so let's get a bit more realistic. Let's say we're shopping and see something that just lights us up. It could be sneakers, a handbag, a car—the item doesn't matter. We are obsessed with how beautiful it is. Everything about it—the shape, the size, the color—all of it is perfect. We think *Ah! I have to have that!* We're experiencing joy and imagining the satisfaction we'll feel when we buy it. The emotions are flowing and the sale is as good as done. Throw in a little logic like *it's on sale* or *I deserve a treat after all I've been through lately,* and you're already handing over your credit card.

The emotion is actually tied to the act of obtaining whatever the object is, not the actual object itself. Which is why when we get home, that thing may end up in the closet or put on a shelf and never used because we have it now, and the emotion is satisfied. The process of obtaining the object caused the joy, not the object itself. This is why so many people are shopaholics.

They become addicted to the positive emotions associated with shopping and just have to keep doing it.

There's a predictable series of subconscious events going on here. First there's a stimulus, then cognitive appraisal of what the stimulus is, then an emotional reaction is felt, which drives a behavioral reaction, which produces a result or function of some kind. All this happens in just a few seconds, and we do it without consciously making a decision about it.

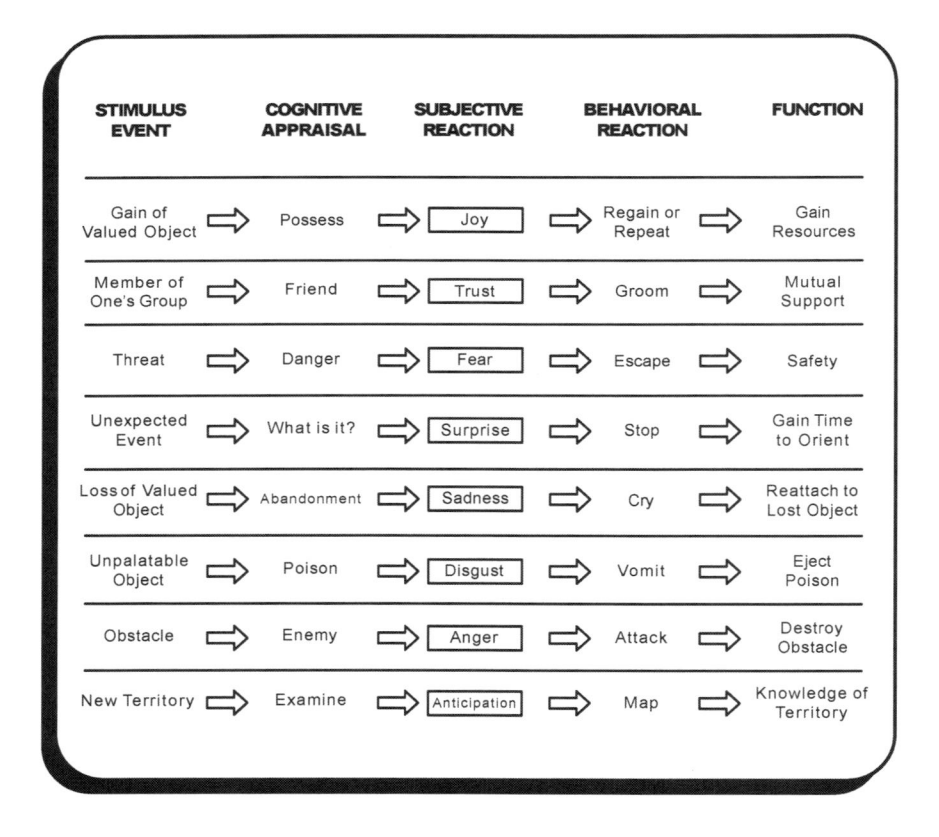

STIMULUS EVENT	COGNITIVE APPRAISAL	SUBJECTIVE REACTION	BEHAVIORAL REACTION	FUNCTION
Gain of Valued Object	Possess	Joy	Regain or Repeat	Gain Resources
Member of One's Group	Friend	Trust	Groom	Mutual Support
Threat	Danger	Fear	Escape	Safety
Unexpected Event	What is it?	Surprise	Stop	Gain Time to Orient
Loss of Valued Object	Abandonment	Sadness	Cry	Reattach to Lost Object
Unpalatable Object	Poison	Disgust	Vomit	Eject Poison
Obstacle	Enemy	Anger	Attack	Destroy Obstacle
New Territory	Examine	Anticipation	Map	Knowledge of Territory

This chain of events applies to just about everything, not just possession of material objects.

Stimulus event: You meet someone on the street who is part of your cycling club.

Subconscious appraisal: This person is a friend.

Emotional reaction: Trust.

Behavioral reaction: Groom the relationship. (High five! "What's up, bro?")

Function: Mutual support and continuation of the tribal group.

If the stimulus event is a threat, the response is even more powerful.

Stimulus event: You hear the sound of squealing brakes and a big crash.

Subconscious appraisal: Danger! Something bad is happening.

Emotional reaction: Fear. What is it? What do I need to be aware of?

Behavioral reaction: Fight, flight, or freeze? Should I run away or toward the danger?

Function: Safety and continued survival.

The emotions can take over your whole body in a threatening situation. Adrenaline courses through your body, and you're at peak awareness and focus. Fear can consume your body and mind. Even people who aren't generally fearful still succumb to this stimulus when other people are involved. If you hear a car crash on the street, you might be alert and wonder what happened. But if your child had just pulled away from you and ran toward the street, you're at a whole 'nother response level!

The function of fear is to help you get to safety, to get back to your comfort zone, away from whatever is threatening you. It's a powerful motivator. And now that you understand how emotions drive actions, you'll be able to create much more effective advertising.

The purpose of advertising is to get large groups of people to take a predictable repeated action, and the only way to do that is to trigger a common emotion using words and imagery that resonate with lots of people.

Emotional combinations

Different emotions have different root emotion combinations. Just like blue and yellow makes green, you can mix emotions to get different variations. For instance, when we feel love, that's a combination of joy and trust. Those two things come together, and we feel that we love someone. We trust them with our future. They bring us joy. When we see the person we love, we're super happy.

Guilt is a mixture of joy and fear. When you feel guilty about something, there's a weird mix of conflicting emotions. On the one hand, you feel you've done something bad and you fear the negative consequences. But on the other hand, the reason you did it in the first place was because it brought you joy, and you feel guilty because you know you shouldn't enjoy bad things. Like having a third massive piece of chocolate cake because it was just so good! You *know* you shouldn't have eaten it. You *know* there will be consequences. But you loved every second of eating it.

Advertising psychology is really interesting when you think about it, because if you know what emotions combine together to get something you want, you can actually create the effect through thoughtful communication.

Have you ever listened to a really powerful speech about something you believe in? It could be religious, political, artistic—doesn't matter. What the person is doing is creating an emotion inside you to make you feel. And when you feel, you never forget. Feelings are stored away as data to be used in future decision making. *Eating ice cream makes me feel good. Great! So in the future, when I see ice cream, I will want to eat it because I know it made me feel good in the past.*

If you think about something bad that happened to you many years ago, you may not be able to recall the details. But you will be able to bring up how you felt. This is why negative events are so traumatic and why PTSD is such a big problem. People in the military go through really crazy experiences that trigger endless amounts of fear and surprise, sadness, disgust, and anger. All these negative emotions experienced at really high levels for extended periods of time overload the system and produce erratic behavior.

It's also why the media and news channels are forever parading bad news across the screen. "If it bleeds, it leads," as the old saying goes. The media's job is not to inform you or to make you feel safe. It's to sell advertising. And it knows the best way to do that is through fear and negative emotions.

I know, you're never going to be able to look at advertising the same ever again, right? That's good! I want you to analyze every ad you see from now on, especially if it works on you. Notice what emotions and memories it brought up for you. And pay attention to the chain reaction of events that happens. What can you learn from it?

Copywriting for pay per call

So, how do we leverage this understanding of human psychology to write better ads and drive more calls? The first thing that we need to do is figure out the words and imagery we want to use in our ads to get people to take the desired action. Here's the process:

- Write down the offer and the audience you're targeting.
- Write down the emotions you want the user to experience and break those down to the core emotions using the chart above.
- Think of a common human memory that evokes those emotions and a story you could tell that would bring up that memory.
- Build a list of words that might trigger those emotions based on common human experience (or at least common human experience in the country you're running the ads in).
- Finally, write a story that brings back that memory using the list of words and images you just made.

This is the ideal way to work on your ad copy. But we also have to work within certain limitations. If you're using Facebook or Google ads, for instance, you can only use so many words in your ad copy or only so many words will be shown. So, we have to create our advertising within the parameters we're given. But in general, try to use stories as much as possible.

OK, let's dig a little deeper into this process with some examples. Obviously, the first thing we need to do is determine who our target audience is. And not just the end user, but the decision maker. If you're selling car insurance to teenagers, it's probably a parent who is actually making the call. So you need an ad that evokes emotion in the parent about the teen potentially getting into an accident and what the effects of that could be.

So, who's your target? How old are they? What ethnicity and gender are they? Are they single? Do they have a family? You're looking for general demographics that apply to large groups of people. What demographic group buys the most car insurance? Probably women, 35-45 years old, and they're buying it for the whole family.

Who's most likely to call for plumbing help? It could be anyone, but the value of the call is higher if the caller owns their home. So, realistically you're probably looking at homeowners over the age of 30. They could be male or female.

Then, you need to put yourself in the mind of a prospect. When do people call for plumbing help? What is that person like in the actual moment? Just imagine: *I'm standing in my kitchen right now. I'm looking out the window, and I'm doing some dishes, scrubbing some dishes, and then, all of a sudden, I reach up to grab the faucet, it breaks off! Water starts shooting out everywhere—like all over the window and all over the wall. I try to stop it. I try to plug it. I grab a towel. I'm trying to shove it in there. I'm like, "Oh my God! Water is going everywhere! What am I gonna do?"*

What am I doing? I'm panicking! I'm feeling all the emotions. Fear! Panic! Anxiety! Uncertainty! Like, "Oh, my God! My house is gonna flood! My neighbors below me are gonna freak out. What am I gonna do? I need to call a plumber."

There you have it. Panic, fear, uncertainty. Those are the things that people are feeling at that very moment in time. So make a list of emotions prospects should be feeling when they have a problem or when they're shopping for something or when they need that something you're promoting.

Next, what memory would make sense and resonate with the largest number of people? What panic-inducing plumbing-related incident might they have faced in the past? Maybe it's the bathtub overflowing. Maybe it's the toilet overflowing. (Then you're adding in a little embarrassment as well ... *It wasn't me. It was like that when I went in there!)*

How many scenarios can you come up with that bring back memories with a bit of panic? Busted pipes? A water hose exploding and soaking your backyard barbeque? Enjoying a shower and then suddenly the water turns ice cold? How about a septic tank overflowing? Come up with lots of ideas so you have plenty of ideas to test.

Once we have our emotions, write a list of words that trigger those emotions. Pay special attention to the target age ranges. A 13-year-old boy uses different language than a 65-year-old woman. They use different words. It's just how we are as humans. About every decade, you're gonna have different words.

Now, in some campaigns, you can communicate the same way to everyone. But, if you're targeting a specific age group like 55+ or 65+, you want to do a little research into words that were popular when they were growing up. If you can do that, you'll be able to trigger emotions that no one else can. Check out these examples:

Ad 1: For the 65+ demographic

> Got a far-out sink giving you the blues?
>
> Our ace plumber, Johnny Pipes, has the skill and know-how to fix your sink in a jiffy.
>
> Call "Groovy Sink Repair" at 1-800-65-PLUMB, and let us make your sink groovy again. Peace, love, and leak-free sinks!

Ad 2: For people in their 20s

> Don't let a broken sink ruin your perfect #KitchenGoals!
>
> Our plumbing hero, Mike, knows how to fix sinks faster than you can say "avocado toast." No more worries about leaks ruining your vintage hardwood floors or Insta-worthy countertops.
>
> Slide into our DMs on Insta (@SinkFixr) or call us at 1-800-20-SINKS to get your kitchen back to being the talk of the town! #NoMoreLeaks

Just search online for popular words in the 1970s or the 2000s. Find out how women and men describe their bodies or their homes. Women use different words than men. People on the East Coast use different words than people on the West Coast. Different ethnic groups use different vernacular. Think about the mindset and the vocabulary of the people you're talking to because you're not writing ads from your point of view. You're writing ads to appeal to *their* point of view. It's not that hard to do, you just have to sit down and think about it. Most people don't do that, which is why so many people fail at creating great ads.

Now it's time to create your ads—lots and lots of them. Approach the topic from different angles with different emotions, some positive and some negative. Write long form, short form, make some videos. Try writing a funny ad and a tear-jerker. Just have fun and create a whole bunch of ads.

And then, no matter what kind of ad you're writing, finish with a call to action. You have to tell people to pick up the phone and make the call. *Call now! I want you to pick your phone, dial 800-899-3726, and call now before it's too late. Don't wait! Call now! We'll provide the relief you need!* Over and over and over again. I know it feels weird, but just keep repeating yourself with the call to action.

Finally, test those ads. We create variations of 10 different ad groups, 10 different images each, 10 different ad types with 10 different emotional traits. Run them all and a couple will just rise to the top. Those are your winners. Put more money behind the winners, drop the losers, and then start all over again trying to find a combination that works even better than your current winners. It's fun if you let it be. Turn this into a game, and try to outdo yourself.

Debt consolidation example

Let's do a debt consolidation example using our wheel of emotions. How does someone who has a lot of debt feel when they think about their debt? People don't like debt. It's associated with negative emotions for most people.

So, what emotions do they feel when they think about their debt? Fear. Shame. Guilt. Anger. Jealousy. Despair. Hopelessness. You can go on and on. The more emotions you can think of, the more angles and stories you'll be able to write.

Now, with debt consolidation, we're not really selling the problem at all. We're selling a solution. But in a debt campaign—or any campaign where someone's in a negative situation and you want to get them to a positive one—you actually need to make them feel things twice. First, you need to present the problem and make them feel pain, and then you need to sell the solution and make them want to attain those good feelings. That's how you generate a lot of profit.

Most ads you see for this niche are really simplistic and don't reach any emotional states at all. *Do you need debt relief? Let us help relieve you of your debt!* They're focusing on the happy emotions. Well, if you don't get them into a state of panic and stress when talking about debt, they're not going to be clamoring for that solution.

All you have to do is take the negative emotions you want someone to feel, and write words that trigger those emotions. Put yourself into the mindset of the person you're trying to influence. Then, literally say those words out loud. Pretend you're in debt, and see if you can actually trigger an emotion in yourself.

Alright, how do I make myself feel stressed? Oh, collectors. Interest, I hate interest. Interest is like being in shackles. Payments and due dates, late fees! Oh my God! I'm never late for anything. Just the word "late" scares the crap out of me. Penalties. Ugh, I'm never going to get out of this hole.

Allow yourself to feel the words. Then the ones that strike the most powerful chord in you, write them down so you can use them in your ads.

Imagine ...

I'm just sitting there watching TV and chilling and munching on some Doritos or whatever, and then an ad comes on. I'm like, *Oh, OK. Cool. Whatever. It's an ad.* I'm watching it. Don't care. Then all of a sudden, my emotions start to shift. *Man, what is with these people? Why are you reminding me of my student loan debt? Come on, guys! I've got $70,000 I've gotta pay back. Oh my God! I'm never gonna get out of this apartment. I'll never afford to have a family and a car that doesn't break down every other week. I'm stuck in an eternal black hole of interest that's making some old dudes rich. I hate this! Give me the solution!*

Once you have someone in that state of anxiety, they want relief—so offer them relief. That's how you get them to take action. You actually have to make them feel the pain. I mean really feel it. They should be squirming around like a kid with a dirty diaper. Just sooooo uncomfortable that they'll jump at any relief you can give them.

We want them to know that if they pick up the phone and call, they'll get relief. Those relief emotions are freedom, courageousness, hope, happiness. You want them feeling powerful, in control, and liberated. That's what you're offering them with the solution. If you're not approaching your ads this way, someone else is kicking your ass at marketing, plain and simple.

Now, you may be feeling like you don't want to sell this way. Or it's unethical or whatever, and I encourage you to get over that. Getting people to experience negative emotions is how this game is played. And if you think about it, you're doing them a favor. What's worse? Feeling a little shame or fear about your student debt or never getting the help you need?

The products and services you're advertising *help people*. So any method you can use to get them to call and ask for that help is a good thing in the long run.

Negative emotion words

OK, let's look at some words and phrases you might write down to evoke negative emotions for debt consolidation. How is your target audience feeling when they think about their debt?

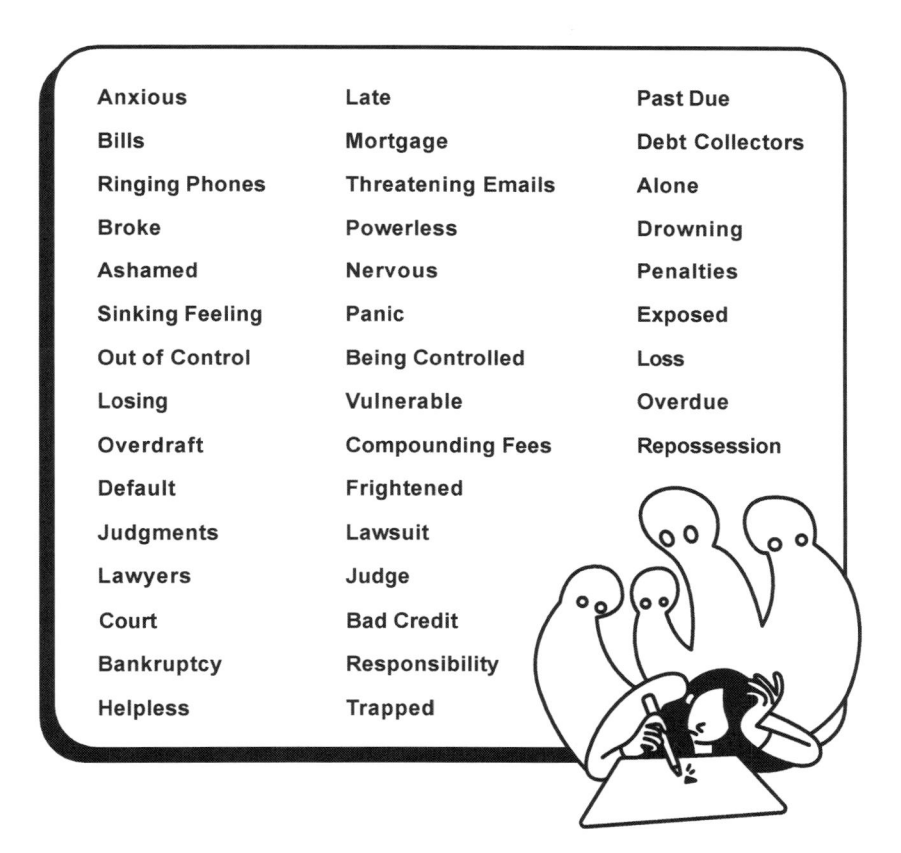

Anxious	Late	Past Due
Bills	Mortgage	Debt Collectors
Ringing Phones	Threatening Emails	Alone
Broke	Powerless	Drowning
Ashamed	Nervous	Penalties
Sinking Feeling	Panic	Exposed
Out of Control	Being Controlled	Loss
Losing	Vulnerable	Overdue
Overdraft	Compounding Fees	Repossession
Default	Frightened	
Judgments	Lawsuit	
Lawyers	Judge	
Court	Bad Credit	
Bankruptcy	Responsibility	
Helpless	Trapped	

Positive emotion words

We also need the other side because we need to communicate that we have the solution.

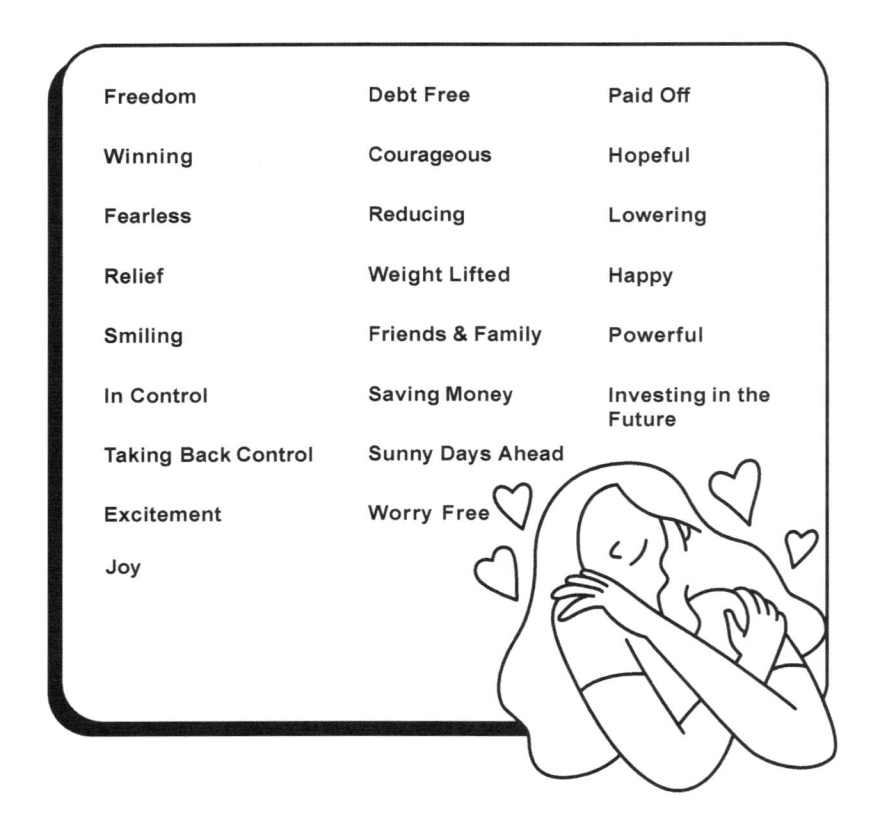

Freedom	Debt Free	Paid Off
Winning	Courageous	Hopeful
Fearless	Reducing	Lowering
Relief	Weight Lifted	Happy
Smiling	Friends & Family	Powerful
In Control	Saving Money	Investing in the Future
Taking Back Control	Sunny Days Ahead	
Excitement	Worry Free	
Joy		

Combine the words to create compelling copy

Once you have the individual words, you're going to combine them into longer phrases and eventually full advertisements. You want those short phrases to trigger something in the target audience.

Some negative examples might be:

- Crushing pressure of due dates, late fees, interest, and penalties
- Phone ringing off the hook
- Debt collectors harassing you nonstop
- In danger of losing your home
- Repo man out to get your car
- Helpless, trapped, and alone

Some positive examples might be:

- Call us, we'll set you free
- Your future is bright, and you have options
- Fight, win, and take back your life
- Celebrate your final payment
- Finally be worry free
- Hold your head high
- Take control and enjoy life
- Our expert lawyers will fight for you

There are more aggressive phrases you can use as well as and softer angles. The only way to know what works is to test them all. You never know what phrase or memory is going to hit the jackpot. It might be a memory of having a debit card declined at the grocery store, shame at having a car repossessed, or the fear of facing a judge in bankruptcy court. Test different scenarios with different audiences until you find what works best.

What does this all look like in actual practice? Here are examples for short-form copy:

Example 1

Is the crushing pressure of due dates, late fees, interest, and penalties ruining your life? Your future is bright and you do have options! Call 800-123-4567 now for a free consultation!

Example 2

Your phone is ringing off the hook but you hide the screen, knowing you can't pay. It's time to fight, win, and take your life back by finally paying off everything. Call 800-123-4567 now for a free consultation!

Long-form copy is created the same way. You just take more of the phrases and combine them on both the negative and positive sides.

Do you feel trapped and alone, drowning in a prison of debt? That feeling of panic from rapidly approaching due dates keeps you from living the life you deserve while faceless collectors and endless late fees suffocate your future.

There is a way out! Reduce your interest rates, lower your debt, and lift yourself up. In no time, you'll be smiling and dancing, living your new debt-free life. Our expert lawyers will fight on your behalf. Call us now for the relief you need to take your life back! The call and consultation are free. Call 800-123-4567 now!

Anyone who's in debt is going to feel something when they read that. They will relate to it. One of those words is gonna get them.

That's it. That's the entire process of weaving emotions into copy for successful advertising, and it applies to every single ad platform that's out there.

The same process works for collecting images that evoke emotions. You want images of the negative and the positive for different points in your ads, so type the emotion you're looking for into the search bar on a stock image website and pull all the photos that make sense. Don't use any images that are vague or generic. You only want the ones that are clearly illustrating the emotion you want.

Pull hundreds of them so you can test them all. I would personally pair 250 images with one piece of copy to find the one with the highest click-through rate. But even if you only create 25 variations and you use the emotion-driven copy, you're going to be way ahead of most people writing ads.

OK, now that you've had a crash course in creating kick-ass advertising, let's look at the different types of traffic sources and how to use them with pay per call. I guarantee you haven't even thought about most of them.

Ready to scream past all your competitors? Yeah, I thought so.

Let's go …

CHAPTER NINE

DIGITAL TRAFFIC SOURCES

When I started there were almost no resources or guides, no forums or communities, and no rules or guidelines. It was the Wild West. We thought it was cool because it was so easy to get started and turn a profit. But it was also a lot harder to find opportunities, advertisers, and partners.

Today it's so easy to find advertisers, advertising platforms, and traffic. The sheer number of tools and opportunities available has created the opposite problem: Where do you even start?

There is a lot more competition today, too. So many people have realized that being an affiliate is possible, and they want that laptop lifestyle, which is totally doable except for the two-hour work month bit—that's bullshit. You have to grind, at least in the beginning, or someone will eat your lunch.

When I got started, if we wanted to advertise on a website, we couldn't just go to an ad network. I had to research the domain on Whois.com, read the terms or privacy policy in the footer, try to figure out who owned that sucker, and then cold email them asking if I could buy a banner placement on their site.

There was no tracking or CPM/Impression-based advertising, so I couldn't just target the people in Montana who were interested in cowboy boots. I had to buy the placement, all or nothing, and then throw away most of the traffic that came to the site or clicked the banner because it wasn't qualified. It felt like panning for gold. Sometimes I would find the motherlode, but most of the time I just burnt a lot of the opportunity because I didn't know what was happening. No analytics, no reports, no nada—just spray and pray, baby.

Today, technology has resolved all those issues. But because there are all these tools, CPM rates and competition has gone through the roof. This is why you really need the right software to take you to the finish line. If a Ringba customer is load balancing with Ring Trees and predictive routing through a supercomputer cluster, and you are trying to manually route calls, you are, as they say, screwed. My client will crush you in ad spend because their ROI will be so much greater.

Digital advertising has come a long way, but in many ways it's exactly the same. And while the mediums or delivery mechanisms may change, text to images, images to video, video to voice, humans to AI, the methodologies and strategies of successful advertising campaigns will always remain the same: Trigger emotions based on memories from the human experience, and offer people a way to satisfy their primal fears, urges, and desires.

In 2021, worldwide online advertising spend was around $522.5 billion. By 2026, that number is expected to reach $836 billion on both desktop and mobile. As of 2022, the internet was considered the most important platform for advertisers, accounting for 62% of total media ad spend. [1]

The three biggest players in the paid ad space online are Google, Meta (Facebook), and Amazon. Google's share of the market has shrunk in the past five years from 31.6% to 26.4% in 2023. Facebook has remained stable, hovering right around 24%. And Amazon has almost doubled, moving from 7.8% market share to 14.6%.[2]

The vast majority of people are accessing the internet and consuming content on mobile devices, including phones, tablets, and handheld gaming systems. Mobile accounts for approximately half of web traffic worldwide. In the fourth quarter of 2022, mobile devices (excluding tablets) generated 59.16% of global website traffic.[3]

With hundreds of billions of dollars flowing through the system, there's unlimited opportunity for you to advertise anything online. The key is to be smart about it. It used to be safe to assume that desktop users were probably working and mobile users were probably browsing in their leisure time. But that's not the case anymore. Since the pandemic, people are now using their devices more than ever—for work and play.

Eighty-three percent of companies have some sort of BYOD (bring your own device) policy in place for workers. And 75% of employees are using their personal phones for work. It's not just convenience driving these policies. The average BYOD policy is estimated to create an extra $350 in revenue per year per employee.[4]

Imagine being able to advertise to people *while they're working*. That never used to happen when we all had desktop computers and worked in office buildings.

Pros and cons of online advertising

The easiest place to start with pay per call is online advertising. It's the lowest barrier to entry and the fastest way to scale your winning campaigns. But, of course, there are pros and cons to consider.

On the pro side, online ads are cheap and you get near-instant analytics to use to make decisions. You can get started with as little as $50 if you like. There's no minimum spend level you have to meet. There's nearly unlimited inventory available, and the platforms will let you play at whatever level you're comfortable with.

You can start, stop, and change directions at any time. There's no waiting around for results. And you can test thousands of ad combinations with different copy, targeting, images, video, landing pages—the works! Those tests can run simultaneously for practically nothing. That means you can figure out what works and then get busy scaling in a matter of days or even hours.

No one can hide what they're doing; all the ads are readily available if you're willing to look for them. Which means it's easy to find examples of winning combinations and model them in your own campaigns. That's why I don't really buy it when people say they can't make their campaigns work. The answers are right in front of them if they just do a bit of research and testing. My personal opinion is that people who can't make it work are basically lazy. If you're willing to learn and test and figure things out, you can absolutely succeed in this business.

Now, because it's so easy to start and succeed online, it's highly competitive. The bigger and better your campaign is, the more likely people will rip it off. There's a difference between using someone's winning ad as inspiration for their own and just plain stealing it. Apparently, stealing is the preferred method for a lot of people in this industry. That's incredibly frustrating when you've put a lot of time, effort, and money into building something truly creative and unique.

Lazy marketers flock to the most obvious traffic sources. They are only interested in what's the easiest and fastest way to win. And there are a *lot* of lazy marketers out there. Even if you find an easy win, it likely won't last very long. Your competitors will copy it to death and then it won't work anymore. It's back to the drawing board. Sooner or later, people playing this game either quit and move on to something else or devolve into being a vulture themselves.

But that's the great thing about driving calls instead of clicks: A human has to answer the other end of the call. What that shows is those buyers don't want unlimited calls; they can't handle the volume, so they are pickier about who they do business with. And if you have a great relationship with your buyers, you can build a moat around your business. You and maybe a handful of others are serving that buyer, not thousands of affiliates driving massive amounts of traffic.

Besides that, not many regular affiliates even know about pay per call, so they don't even know what to do with your style of ads. Their goals are completely different, and your campaigns won't work for them. That's why I love pay per call. It's so much harder for people to rip you off.

Even so, in the digital ad space, you will have people trying to copy you. You have to decide if it's even worth your time trying to compete with all those people. If you're creative and think for more than five minutes about what you're trying to do, there are all sorts of alternative ways to advertise—methods that don't have a ton of competition and vultures just waiting to poach your campaigns. And frankly, thinking and being creative are a lot more enjoyable anyway. I get to have fun *and* gain a competitive advantage in the industry. That's a win in my book.

The final negative I want to point out is that unlimited opportunity also has a downside. Sometimes it can be paralyzing to look at so many ad platforms and creatives and format options. Facebook or Google? Short video or long? Relationship or transactional? Text with images or without? Not to mention all the millions of things you could be promoting. Liké, how are you supposed to choose one way to go?

Here's how: Pick a vertical and a platform. Put one foot in front of the other and learn that platform until you've mastered it and are making great money. Then, if you want to, choose another vertical or platform. The truth is if you can master one, you can master them all using the exact same process. But if you're distracted by a thousand shiny objects, you'll never get anywhere.

How online advertising works for pay per call

I'm assuming you've already chosen a vertical and have a buyer to send calls to. That's the first step. The next step is choosing your digital traffic sources. What playground do you want to hang out in? Google? Facebook? Bing? Pinterest? Don't discount the lesser players. If you can make a campaign work in a less competitive platform, that's a huge win for you. Just pick one and dedicate yourself to learning it inside and out.

Because Ringba is one of the biggest platforms in the pay per call space, I have insider access to what's working in this industry. I see the stats every day for every vertical imaginable. You want the secret? The real scoop?

Everything is working.

All of it.

For every way you can advertise online, we have customers making millions using that method.

Really.

So stop fretting, OK? I get so tired of people saying, "Google doesn't work anymore." Give me a break! Of course it works. Every traffic source works for almost every type of campaign as long as you take the time to figure out *how* to make it work. Don't be lazy.

So, which one should you choose? I recommend starting with a method you've used before. If you've never advertised online, then start with the biggest and easiest—either Google or Facebook. They're the easiest to get started with, especially at lower budgets.

Once you've got your platform, next you build your creatives. These are all your ads, landing pages, emails, copy, images, and videos—whatever you need to actually create and run your campaign.

So far so good, nothing really new here if you've run traditional affiliate campaigns before. But next is something new, it's time to set up call tracking. As I've said from the beginning of this book, if you don't have your own call tracking service, you're going to have a really hard time winning. You have to be in complete control of your own destiny, which means being able to route and load balance calls between multiple buyers as well as doing your own quality assurance and fraud detection.

At this point, you've got your buyers, your platform, your creative, and your call tracking. It's time to rock 'n' roll. Set up your campaign and build lots of tests. I'm talking hundreds! I used to start with at least 250 multivariate tests just for the ads alone. I wanted to see what combinations of text, images, video, and placement would get the highest click-through rates.

How do you split test hundreds or a thousand creatives at a time? Well, that's pretty easy. First, you get a thousand creatives. Second, you upload them into an advertising platform. And third, you turn it on and let them start to get traffic. I don't mean to be flippant here, but that is literally the process. Most experts espouse testing just one variable at a time so you know exactly what changes. But that really applies more toward landing-page optimization and not creative testing. At the end of the day, the creative is far more valuable than landing page optimization.

I'm not saying you should avoid your landing page, but if you have great creative and a solid offer, it really doesn't matter what your landing page looks like. People will figure out how to give you money. On the other hand, if you don't have strong ad creative, no one will see your landing page at all. So start by testing a ton of creative options. Then once you're getting lots of traffic, start testing your landing pages.

I figured this out—surprise, surprise—when I was in my parents' basement trying to make things work. It's funny how desperation really sharpens your sword. I was so frustrated because my creatives were burning out. Some were successful, some weren't, and I couldn't figure it out because I was only doing a few at a time. So, I decided to split test a ton of options to find the best one that would crush it. The problem was I didn't have much money. How could I test?

This wasn't in the pay per call space, it was an affiliate offer for a dating site. I decided to play catfish for a while and set up some fake profiles just for the purposes of getting some free testing. I took a whole bunch of pictures, and I put them on hot-or-not.com. As people randomly voted on my pictures, I figured out which pictures got the best results, the angle, the smile, everything. I optimized until I got to a 9.9. Then I took that photo and ran it in a Facebook ad. And believe it or not, it actually resulted in a high-converting creative, and I made some great money off of it.

These days, of course, video ads are the way to go, so I recommend you start by coming up with a video brief. List everything you want in the ad, then hire three different people to create three completely different ads. (This doesn't have to get expensive. You can hire people on Upwork or Fiverr who will work fine. Or better yet, make the video yourself. iPhones are incredible tools.)

Then run those three ads and see which one works the best. Maybe the video with someone eating a hot dog does really well. Great, so you run with that. Go back to your freelancers and have them each make 10 different versions of a guy eating a hot dog. Then do 10 with a woman eating a hot dog. And 10 more with kids eating hot dogs. Hot dogs in the kitchen, on a picnic, people throwing hot dogs to their dogs—as many ideas as you can think of. You can run different soundtracks and voice-overs. Just get as many variations as you can. They should all be promoting the same thing and have similar calls to action. Now you have several hundred different ads using a basic idea that you know works.

Then you test them. Run them all on Facebook or YouTube or wherever you like, just to see which one performs the best. And you only need a few thousand impressions to see which will get the highest click-through rate.

I'm aware this is an insane amount of editing, and that's why you need to have people help you with it. But if you test a thousand creatives, you will find a whole batch of them that absolutely crush it at super high margin. The thing is, most people are not willing to do this much work. They prefer to rip off someone else's ad and ride those coat tails. But the most valuable thing you'll learn by going through the process of creating hundreds or a thousand creatives is to know what makes a great ad. You'll figure out what to look for, what performs better, what audio works better, what video, what copy, all those different pieces of the puzzle. That's education you'll never forget. And the next time you run an ad, it will cost next to nothing because you will have an innate sense for what works.

I did all of this myself because I didn't want to wait for contractors to do it for me. I just wanted to grind out the iterations until I knew what worked. But you can work with what you're comfortable with. If a thousand is too much, start with three. Then scale to hundreds as fast as you can.

Is that a lot of work? Damned right it is. But when I win, I win big!

And guess what? You have it even easier than I did because platforms like Facebook will use AI and all their own data to put together the variations for you. All you have to do is load in the different assets and let the platform figure out the variations most likely to win. You can't get easier than that.

Once you figure out the type of ad that's working, you're going to create even more tests. Like 500 to 1,000 more. The only way to figure out which ads have the highest click-through rates is to test the living shit out of a ton of combinations. You're just grinding away. It's not glamorous. None of the gurus will talk about this part because most of them don't do it—at least not to the extent I want you to.

I still remember when Facebook advertising was a new thing, sitting in my parents' basement churning out thousands and thousands of ads. And I made a ton of money because once I found out what worked, I could scale it. It's all about how much work you're willing to put in up front. That's the game. You don't have to be a rocket scientist. You just have to work your ass off in the beginning.

You're never going to stop testing. Ever. Once you find something that's profitable and wins, you're not done. Some people say, "My campaign is running. I'm making money. Awesome." And then they stop. Great. But someone's going to come along and take it away from you if you don't keep working. Or the ad will fatigue and stop working on its own. That's just the nature of online marketing.

You need different angles, different landing pages, maybe different niches and demographics. You're going to review your results daily, hourly, maybe to the minute. If you're spending $25,000 a day on media, you better have someone watching your campaigns around the clock. No, you don't have to be on call 24/7, but someone needs to be paying attention.

You're never finished. Testing and monitoring—that's your career. It's fun to hit the refresh button and watch the money roll in. But pay attention to the details that matter. What ads are working? How can you optimize them? What buyers are converting best? And honestly, once you understand your job, you can really start having fun. With persistence, you can *own* entire verticals.

Click-to-call advantage

I've said before that a phone call has the highest buying intent of all advertising actions. When someone picks up the phone, they're serious. And they don't want to mess around with anything as mundane as dialing a number. They just want to make one click and have the call go through. Fortunately, we have that capability. They're called click-to-call links, and they're really simple to create.

Here's what a normal link looks like in HTML. You just set your href to the URL you want the user to go to, like this:

<a href="url">Link text</a>

A click-to-call link is very similar. But instead of the href, we're gonna put T-E-L colon, the area code, with the country code and the full phone number. Like this:

<a href=tel:YOURNUMBERHERE>Call now
YOUR NUMBER </a>

Click-to-call links power the vast majority of the pay per call industry. They work anywhere people can get access with a mobile device, including landing pages, email, and ads. Mobile users click or tap a click-to-call link, and a dialogue pops up on their phone that allows them to call that vendor immediately. Memorize how to write this simple code. You're going to be using it a lot.

QR to call

The first time I saw a QR code used for pay per call, I was watching infomercials with Harrison in my favorite room at the Bellagio hotel in Las Vegas. It's a giant circular room that's right in the middle of the rotunda looking out over the fountains. Anyway, we were walking in circles, talking business and this ad came up on the TV that said *scan this code to go to our website.*

I thought, *Huh. I wonder if you can put a click-to-call link inside a QR code.* So I grabbed a laptop and embedded a click-to-call link inside a QR code. Harrison scanned it with his phone and my phone rang! It was like we'd discovered fire. We could just tell people to scan the code, and they wouldn't have to remember the phone number. We could use as many different tracking links as we wanted because we're just generating QR codes.

We had this epiphany moment in the middle of the Bellagio room, and then, all of a sudden, the fountains went off and the music was blaring! Like the gods of Vegas were congratulating us on figuring out something so cool.

People aren't using it as widely as I think they should be. The pandemic trained us all to use QR codes all the time, so people are used to them now. You can use QR codes anywhere—in direct mail, in video, even on an ad on a movie screen. It's pretty amazing.

Types of Online Advertising

Remember earlier when I said all the online marketing methods work for pay per call? It's true, as long as you're willing to make an effort. But that can still leave you frozen in your tracks just trying to pick a method to focus on. So let's look at the different methods briefly.

Paid search is most often referring to Google AdWords. It's been around forever and it actually has its own call-only ad platform as well as ad extensions and site links where you can add phone numbers. A lot of people I know start in paid search and many never leave because it just works.

Search is highly effective because the consumer is signaling an intent. They are initiating the search for something, which means they're looking for an answer. Depending on the vertical, they may be looking for an immediate need. When someone calls for a plumber or a tow truck, chances are they aren't screwing around. You're catching them in their moment of need.

For instance, they may search, "I need a plumber," or "plumbers near me." Placing ads on key phrases like those are going to be highly competitive, so you can get more creative with phrases like "My kitchen is flooding." There will be lower search volume but also a lot less competition. Your job here is weeding through all the available search possibilities to find the ones with the highest volume and lowest competition.

You also need a big list of negative keywords to stop ads from being shown to people who are not signaling the correct intent. If someone is searching for YouTube videos on how to fix a sink, you don't necessarily want to pay for an ad to show up. You should always ask your networks for lists of their negative keywords for every single campaign because sometimes it's more important to have a good negative keyword list than it is to find positive keywords.

Google call-only advertisements are really the way to go with search for pay per call. The platform is exclusively reaching people on mobile phones, and all the ads initiate a phone call when clicked. They don't give the user an option to go to a landing page, only make a call. Also, Google pays close attention to your budget, your CPA, and your hours of operation, which is critical when working with call centers.

Some people are getting into trouble with call-only and getting their accounts banned. Nine times out of ten it's because they don't know the rules. Google's call-only platform is not the same as AdWords, OK? There are different requirements, including having an official LLC or DBA and disclosing the name of the business when someone calls. Terms of service change all the time, so make sure you read and understand them. This platform is too powerful to risk having your account shut down.

Social media for pay per call is still a practically unlimited resource in the industry. Social is not intent based. It's interruption based. Your ads are interrupting people and trying to get them to take a different action, so you have to inspire them. You have to actually sell them on what your product is and why they need it. It's a different animal and a different approach.

Social advertising for pay per call requires more of a storytelling format. And the story has to resonate with the consumer. A plumbing ad probably isn't going to work on social because someone scrolling through cat videos probably isn't worried about their pipes bursting.

But a retirement home ad that leads to a helpful article for adult children of seniors? That's a much better fit. Anywhere people need help and understanding gives you the opportunity to win their call with empathy and solid information.

The best format that everyone in social media uses is called an advertorial. An advertorial is a hybrid mashup between an ad and an article (called "editorial" in the print world). The way it works is you write a really nice educational article about a topic, complete with images, videos, testimonials, and several calls to action. You post this on a landing page and run social media ads to send traffic to it. Make sure your ad style and tone match what people will find on the landing page. You have to meet the expectation you set in the ad.

Also, make sure your landing page works on all mobile devices. You can get software that tests mobile compatibility for you, or you can just buy all the different types of phones. This is one major area where pay per callers trip up. Make sure people can see your advertorial when they click on your ad!

Sometimes people click links in advertorials and don't realize they're making a phone call. So add a little phone icon on your button or next to your links. Breaking expectations is a sure way to break trust, and you don't want to do that.

Push notifications are some of the lesser-used methods where people are winning in pay per call. Push notifications are short messages that notify a consumer of all sorts of things like news, weather, traffic as well as sales going on at a particular store. They're sent by apps and websites that a consumer subscribes to.

The beautiful thing is the process happens whether people are browsing or not. The notifications just show up on their phones at certain intervals, they click the message and go to the promoted landing page or offer. You can put a click-to-call link right in the notification, or you can use a normal link to drive a user to a landing page that's designed for pay per call.

While there's a lot of opportunity here, there's a lot of panning for gold involved. If you can't afford to lose $500, I don't recommend that you start with push notifications because it's gonna take a lot of testing to land on a winning strategy.

Email has the highest value of any type of communication. It always has. You're not competing with distractions in a newsfeed. You're directly inside their inbox. The problem with email for affiliates has always been the links. For the longest time, having links in your emails would cut deliverability rates.

But there are no links in a pay per call email, just a phone number. Email companies have no problem with you embedding phone numbers, so the deliverability of the email goes way up. On iPhones, you can even put the phone number right in the subject line to automatically initiate a phone call.

You can rent a list, you can build your own list, you can borrow someone else's list—it doesn't matter. If you can write a solid email that persuades people to call, you're golden.

Now, I know what you're thinking: *Sweet! I'm gonna smash out 20,000 emails and watch the money roll in.* Yeah, about that ... remember, with pay per call we're dealing with call centers and actual people on the other end of the line. A *limited number* of people. If you send out too many emails at once, you're going to overwhelm the call center and flush your ROI down the toilet. Even when you have your own call-tracking platform and can load balance calls between several buyers, you still have to be careful not to overwhelm them or have people calling when the center's not open.

You need to control delivery. Send small, highly targeted, highly segmented email drops until you understand the math—how many emails you send versus how many phone calls come in. Then pace out your delivery throughout the day based on the capacity so that you can get as many of those phone calls as possible answered.

Just like with ads, you want to test lots of copy, imagery, and subject lines. Put phone numbers in the subject line, or don't. Send them to a landing page or go straight to click-to-call. Use pictures or testimonials. Be creative and test lots of options and combinations so that you can figure out what generates the most phone calls from email.

Native advertising is very similar to social. This is when ads show up on websites, like news sites and blogs. The ads may be designed to match the topic of the content on a particular page, or the ad may actually *be* the content. If you can make your interrupt campaign work on social, you can make it work on native. You target people by what they're interested in, tell them a story that makes them want to check out your products.

Native landing pages require long-form sales copy that tells stories and elicits emotions. You'll probably want to use an advertorial because it's emotion-driven marketing. You have to make them *feel something* to get them to call.

You just have to figure out what ad combinations get clicks and then tweak and tune your landing page for native. This is a multistep optimization process, whereas with search ads, you're only optimizing for keywords and clicks and calls. You're not messing with landing pages, for the most part. With native, you're going to test ads, images, videos, copy, your landing pages, and what websites the ads get placed on. So, it's going to take time, but there's a huge amount of traffic available.

SEO works beautifully with pay per call when done correctly. You can build all sorts of properties, drive traffic organically with well-crafted

SEO, and include prominent phone numbers on those sites. Some of the most successful people I know in pay per call make a lot of money with SEO.

You can't just slap up a $5 blog post, add a phone number, and expect it to work. You actually need to put some effort into creating the landing pages. The demographic for SEO skews older, and these people are actually reading to understand something. They want the real deal, not some bullshit article that doesn't actually say anything. You're not going to get the call unless you provide legit information.

If you don't want to hire a writer, you can use video content that's transcribed and cleaned up to make it readable. For me, it's easier to talk for an hour than to write for an hour. So we make lots of videos at Ringba, transcribe them, clean them up, and post them on our website. Google eats up the keywords in the text, while most of our visitors just play the video.

Another popular method is to use AI writing tools to auto-generate articles on any topic under the sun. I don't recommend using these articles as they come out of the chatbot directly, because Google and other search engines have caught on and will penalize you for using AI-generated text. However, you can take the first draft and have a writer or another software program rewrite it to sound more human. AI is a huge time saver.

Now, if you're going to invest the time and money to do SEO, you need to create a brand. That means purchasing a *dot com* domain name. Don't fool around with this. I know .io is cool or whatever, but no one trusts it. People trust .com extensions. So just use them. Buy a memorable domain name. It took me an hour to find Ringba.com, but I paid $9 for it, and now I have a great brand name.

You also want to invest in solid design logos and page-speed optimization. Your whole website should look clean and professional. You don't have to spend a fortune, but don't cheap out on your web design. Find an awesome contractor who can optimize for mobile and for load time so that Google likes your site and will send it more traffic.

You need to test everything you're doing on all devices. Use a cross-browser testing service to be sure. I would say more than 90% of affiliates aren't doing this, and it's a major reason they fail. If you have errors on your site, people don't trust you, they can't read the content, and they don't call. They don't do anything. So make damned sure your site works. Everywhere.

Link building is the final piece of the SEO puzzle. Be really careful with shady link-building services, all right? They're shady for a reason. If they're promising the world to you, it's probably not gonna work. Manual outreach is by far the most effective. It takes time, but it's worth it. If you're looking for a shortcut to acquiring backlinks, buy a domain name that's been around for a long time. You can do that on sites like Flippa. It might cost a couple thousand dollars, but the site is already indexed and has some backlinks, so you get a leg up.

SEO is a long-term play. It can take time to win with this method. But in that time, you're building a solid, trustworthy brand that will only enhance all your other traffic-generating efforts.

Local search is another interesting option for digital traffic. Google Business Profile (formerly Google My Business) is the largest platform for local search, and some people are finding success setting up listings for local areas within certain niches. They monitor the phone calls that come in and then route the callers to the appropriate buyers. Anytime anyone searches on Google for a local business, like pest control in Poughkeepsie, they see a host of options inside the business search results, so people are used to seeing and interacting with the platform.

Local search platforms do have unique rules that you need to be aware of, and they change every now and then. So make sure you check out the current rules before spending your time building something that may not last. One of the rules is you need to have a legitimate business set up, and you need to identify the business immediately when people call. So, it's a good idea to have an IVR set up that says, "Hey! Thanks for calling Poughkeepsie Pest Control. Please hold for the next available agent. We'll kill anything."

You still have to play the SEO game to get your listings to show up at the top of the search results. If you don't want to do that, you can also buy ads on the local platform and just bypass the SEO.

There are about a million nuances to each digital platform, and those are beyond the scope of this book. But just know that digital traffic is the cheapest and simplest way to get started.

Now, are you ready to leave the massively competitive digital landscape and dive into the clear blue waters of *offline* media? I think you'll find it incredibly refreshing.

Let's take a look at print advertising next. Keep reading ...

CHAPTER TEN

PRINT ADVERTISING

When I started diving down the rabbit hole of hobby and special-interest publications, I had no idea how far it went. I mean, how many different hobbies can there possibly be? My friend, there are thousands! I remember hitting our local Barnes & Noble to do a little research and being blown away by how many magazines they had on display. There were probably a dozen different publications each on gardening, quilting, guns, music, politics, art—and that was just the mainstream magazines for the largest markets. When I took my search online, I realized there were so many magazines, newspapers, and printed newsletters that you can only get by subscribing and having them delivered in the mail.

It might surprise you, but print advertising is absolutely massive. It encompasses several different mediums, including newspaper, direct mail, magazines, local circulars, flyers, and directories. Print ads drive *trillions* of dollars in global sales.

It's in every country in the world, and the best part is that most marketers have forgotten all about it. There are very few pay per call ads in print. If you figure out how to do this well, you can create a sustainable long-term business with very little competition.

The word on the street is that newspapers are dead, only suckers pay for direct mail, and nobody reads magazines anymore. But that's just not true. And the people who do read those periodicals are diehards. They *really* read and pay attention to the articles and the advertising in print. And just about everyone reading the newspaper has a phone in their pocket, so the desired action can happen instantly.

According to research done in 2021, newspapers reach more than 25 million people in the United States every day. That's almost 10% of the adult population. Every. Day. And people who read newspapers are typically in a higher income bracket, so they have more purchasing power. Even with advertisers moving online, newspapers still bring in over $9.6 billion in advertising revenue every year.[1]

You might think only older people read print periodicals, but that's not true either. Close to 95% of people under the age of 25 read magazines[2], which is pretty surprising considering that they're typically glued to their phones. Magazines have a tactile feel to them. They're fun to read because you can actually hold them. Some magazine publishers have been around for over 100 years. They're not going anywhere, even if their readership is shrinking a bit.

Marketers tend to dismiss direct mail, too. But if you target the right audience, those little postcards and letters can produce returns of more than a thousand percent. Direct mail typically has 37% higher response rates than email.[3] Probably because we're so inundated with email these days that the physical "snail mail" has become different and interesting—just like 30 years ago when email was the new and interesting break from all the direct mail we received constantly. Sure, email is cheaper and easier to deliver, but direct mail is still a really exciting opportunity if you're willing to learn a different medium.

Print ads may or may not be in color, and they're not clickable. But you can find highly targeted, cheap placements pretty easily. If you can get them to work, and they're only a couple hundred dollars a month, you've got a great set-and-forget campaign. You just keep buying ads and running them week in and week out.

It's also simple to test and tweak advertisements in print. Depending on how often the publication is printed, you may be able to test ads monthly, weekly, or even daily if you wanted to. It's as simple as writing out a new ad, swapping out photos (if it's an image ad), and submitting it to the publication. To test which ads are working best, you simply use a different tracking number for each one.

Very few marketers even think about print advertising because it's old school. It's not sexy. But I don't care if it's cool or not. I never have. I only care about the opportunity and whether the financials make sense. If I can make money with it, and there's no competition, that's where I'm going to go. The entire internet could crash tomorrow, but if you're diversified into both traditional and digital advertising, you'd be just fine.

If you've ever had trouble with digital marketers stealing your content or ad copy, print should feel like a dream. It's highly unlikely your competitors will ever even see your ad in *Retirement Homes Weekly— Dallas Edition*, let alone want to copy it. Most marketers stick with digital because it seems like less work. But as I mentioned earlier, if you can get a few ads working in print, you can just set it and forget it. It's a business that runs on autopilot.

Way back in the dark ages when websites were brand new, companies scrambled to figure out how to monetize this new medium. Since the closest thing they had was magazines and newspapers, they modeled the advertising after print. Funny, huh?

Consequently, anything you can do on a website with advertising, you can do in a print publication. Great big banner ads or teeny tiny classifieds, inline, headers, footers, full-page, and paid advertorials. With a little creativity, there's a lot you can do to take advantage of a practically forgotten medium. Just think about what you might do with a text-based digital medium like a website, blog, or news site, and then find the print equivalent.

Print advertising makes you think differently, which can help your digital ad game as well. And I have to tell you, it's really fun. If you can develop patience and have an offer that aligns with a publication, you'll win in the long run. And that's what pay per call is all about—the long game.

Working with print publications does have its negatives. It can take a long time to distribute your ads. If you're working with a quarterly or monthly publication, they will need your ads weeks or months before they ever get printed. For daily or weekly publications, you can get your ads distributed faster, but they fade faster, too. An ad in a daily newspaper is most effective that one day, then the longer time goes on, the less chance anyone will even see it. So, you have to run ads day after day to get solid data on whether they're working or not.

There's a lot less room for ads in a physical medium. Digital feeds have seemingly unlimited space for advertising. But a newspaper or magazine can actually sell out of ad space, and you just have to get in line to wait for an open slot. It could take six weeks to effectively test three different ad spots. The same test in digital marketing takes milliseconds. Most marketers want results immediately, and if they can't test 500 ads by lunchtime, they're not happy campers.

Another downside is you're going to need a lot of phone numbers if you want to run a lot of ads. Think about it, if you want to hit the classifieds in 500 major metropolitan areas and those have different daily, weekly, and monthly newspapers and magazines, not to mention niche publications, you may need thousands of phone numbers to test and scale the business.

You really can't play the print ad game without a tracking platform, it's just too complicated. You're going to make a few phone calls, and before you know it, you have an army of account reps to talk to at all sorts of different publications. The industry as a whole is in a downward trend, so these reps are hungry for your business. That means they're willing to take the time to educate you on how to play the game. Listen to them! They've been around the block a few thousand times.

The good news is phone numbers are relatively cheap, but getting them does require some setup time in your tracking platform so you can keep track of every single placement. But what you get in return for that work is granular statistics that tell you which ads are winning, which ones are losing, and how to build your business in a relatively low-competition environment.

The last downside is once an ad is in print, it's final. You can't pause the campaign or change the ad midstream. You can pull your ads and end your contract, but that won't stop your ad from working if Aunt Mabel is looking through a two-month-old magazine at the hairdresser. The ad stays active until every issue of every print run has been disposed of. Calls can still keep trickling in, and you don't want to lose them just because you pulled the ad.

So, how does print advertising actually work? Let's walk through the process. I think you'll find it's similar to any other form of advertising, you just have to dig a little deeper to find the information you're looking for.

Know your target market

No matter what advertising medium you're using, the first step is always *knowing who you're trying to reach.* If you're working on a campaign for Medicare, you'll probably want to attract different people than a campaign for adventure travel. There may be some crossover, but the goal with pay per call is to reach the largest audience possible with the fewest ads.

Begin by creating an avatar for your campaign. Who are you trying to attract? How old are they? Does gender matter? Does location matter? Do they have a family? How much money do they make? What's their education level? What hobbies might they enjoy? Where do they vacation? Do they drink alcohol? Are they vegetarian?

Keep asking questions until you get a well-rounded representation of your audience. Then it's time to start looking for publications that the audience might read. Use the next few pages to get started.

Campaign Avatar Profile Creator

Use this worksheet as a framework for creating avatars for any type of campaign. Answer the following questions to develop a detailed profile for your ideal customer avatar:

Demographics:
Age: _____
Gender: _____
Location: _____
Marital Status: _____
Education Level: _____

Psychographics:
Occupation: _____
Income Level: _____
Interests/Hobbies: _____
Values/Beliefs: _____
Goals/Ambitions: _____

_____

Challenges and Pain Points:
What are their main challenges or pain points?

_____
_____

Motivations and Desires:
What are their primary motivations and desires?

_____

Media and Information Consumption:
Where do they get their information? (e.g., social media, websites, magazines)

_____

What social media platforms do they use?

_____

Buying Behavior:
How do they typically make purchasing decisions?

_____

What factors influence their buying decisions?

_____

What objections might they have to your product or service?

_____

Preferred Communication Channels:
How do they prefer to be contacted?

_____

What types of content do they engage with the most? (e.g., videos, articles)

_____

Customer Journey:
At what stage of the customer journey are they most likely to convert to a pay per call lead? (Awareness, Consideration, Decision)

_____

Goals for the Campaign:
What specific goals do you want to achieve with this campaign?

_____

Unique Selling Proposition (USP):
How can your product or service address their pain points and fulfill their desires better than your competitors'?

_____

Key Message:
What key message will resonate with this avatar?

_____

Conversion Strategy:
What specific strategies will you employ to convert this avatar into a pay per call lead?

Value Proposition:
What value can you offer to entice them to take the desired action?

Engagement Triggers:
What engagement triggers or incentives can be used to grab their attention and maintain their interest?

Objection Handling:
How will you address common objections and concerns of this avatar during the campaign?

Next Steps:
Based on this profile, what are the next steps for this campaign execution and optimization?

This Avatar Creation Worksheet is designed to help you gain a deep understanding of your target audience, which is essential for creating effective pay per call marketing campaigns. Fill in the blanks and use the insights to tailor your marketing efforts to your ideal customer persona.

Research the playing field

Once you know who you're trying to reach, the fun really starts. It's time to hit the bookstore! OK, I know you're probably rolling your eyes at me, but your local Barnes and Noble is a goldmine of opportunity for market research. You can spend hours just browsing through dozens of racks of niche publications from *Knitting Today* to *Guns 'n Ammo*. Yes, you can do all your research online. But it's more effective if you can hold the magazine, flip through the pages, and look at how the ads appear on the page.

If you want to dive really deep into a certain publication, call your local library to see if they carry it. If they do, you may be able to get your hands on back issues, which can help you learn all sorts of interesting information about who advertises there and how often.

The last step is to head online and look for more opportunities. Not every publication is for sale in a bookstore or newsstand. Thousands of tiny niche and industry journals are only available by subscription. They may be smaller publications, but that means their advertising is less expensive, and their readers are more likely to pay attention to and act on your ads.

When I was preparing the Pay Per Call Masterclass a while back, I remember getting so excited about how many micro-niches there are. I started with *American Angler* and just kept moving through the alphabet, digging into all sorts of different niches.

The publication's media kit is where you find all the demographic and target market data. At a glance, you'll be able to tell the age, gender, income levels, and more about that publication's readership. Obviously, you want their demographics to line up with your target market. If the average reader is 35, and you need 65 or older for a campaign, you can cross that one off and move along.

The media kit will also have all the circulation and readership numbers. Circulation is how many copies are printed and distributed. Readership is an estimate of how many people actually read each issue. Circulation is usually verified by a third-party audit, while readership is an educated guess. Sometimes actual readership can be much larger than the circulation because institutions like universities and libraries order one copy to serve many readers.

You can find media kits in a couple of ways; look in the footer of the publication's website for the words "media kit" or "advertise with us." Or you can just Google the publication title and "media kit" to get a direct link.

If you can't find it in those ways, just send an email through the publication's contact page on their website with the words "Advertising media kit" in the subject line. That will get their attention.

If you're anything like me, you'll come up with dozens of great ideas just flipping through media kits in different niches. I recommend finding five great publications that match your target market and then start calling their advertising departments. Their phone numbers will be prominently displayed throughout the media kit. Advertising is their primary revenue stream, so they want you to call them.

Negotiate your price

Advertising prices are usually based on the size of the ad—full page, half page, quarter page, or smaller. Sometimes a publication will also have its own classified ads in the back. The bigger the ad and the more prominent the placement, the more expensive it will be. Also, the bigger the publication is, the more expensive the ads will be. *Rolling Stone* magazine ads are likely going to be much higher than *Drum Kit Monthly*, even though they have some audience overlap.

I always recommend that you get your feet wet with small ads first. That means an eighth of a page or a sixteenth of a page, or even just a four-line classified ad. Test out the print format and see how things go. Once you get good results, then you can scale with bigger ads and better placements.

I know this sounds like I'm contradicting myself. With digital ads, my method is to test fast and scale faster. But that's because digital gives you instant feedback, and you can stop and change campaigns any time you want. That's not the case with print. Here you want to test small and scale fast once you know you have a winner.

Now, the rates in the media kit are what's called "rack rate," which is the standard rate they're supposed to charge. But here's the funny thing about print—once the issue is printed, that's it. They can't sell you an ad for last week's issue. For them, that means they have to sell as much advertising as they possibly can in advance. For you, that means you can often get special deals. If they're having a hard time selling the inside back cover of their magazine, they're going to discount it for you. In their eyes, it's better to collect some money for that space than no money.

You never want to pay rack rates if you can help it, which means you're going to negotiate with the salesperson selling the ads. They will have the upper hand at first. You're most likely gonna be dealing with a weather-beaten veteran of print advertising. And if they're cool, maybe they help you out. If they're hungry, maybe they will help you out. But typically, if you're new, they have the upper hand because they know what the rules are and you don't—yet.

There are lots of ways to negotiate print ads. I'm going to share the simplest way here. But if you want to explore the ad negotiation and contracts more deeply, hit the Pay Per Callers forum and watch the Pay Per Call Masterclass.

The simplest way to negotiate ad rates is to call the advertising department and tell them you're looking to test out a concept *for your boss*. Tell them you're trying to prove that print ads still work, and once you prove it, you can get a bigger budget. But for now, you're looking for their best introductory price.

You can phrase it like this:

> We're a digital marketing agency, and my boss thinks print is dead. I think it'll do really well. If I can convince my boss to let me run a test with you, then I can come back and do a multi-month with a better placement. But I have to show him results first. Can you help me out? I really want to prove him wrong.

It's that simple to request some introductory rates. Just frame it like you and the ad rep are a team and your boss is the enemy. Make it a nice, friendly, collaborative relationship. If it's getting close to their sales deadline, you may even be able to get what's called a remnant rate, which is basically a fire sale on all the ad space that didn't sell. One of my favorite strategies is to find out from the rep when the deadline is for the next issue. Then call the rep the day-of and say, "Hey I've got an ad ready in all your standard sizes, is there anything I can buy on the cheap if I can get you creative in the next five minutes?"

Make sure you have those ads print ready so you can just email them over right after the call. If you do this with a bunch of different publications you will find high margin wins or cheap ways to test your ads.

Don't argue over their rates, just see if there are added benefits they can add on to sweeten the pot. Like, say, an extra ad in their email newsletter or a few shoutouts on social media. You want a long-term relationship with your ad rep because they can often slide you some sweet deals when they have extra ad space. There are all sorts of perks they can hand you if they like you.

Create your ads

So, how do we actually produce the print advertisement? This may seem a little scary if you haven't done it before, but believe me, it's not at all complicated. In fact, it's pretty close to child's play. You just have to know what you want and what the publication expects from you.

The first thing you want to do is get the specifications and deadlines from your ad rep. They will tell you what size to make the ad, how to format it, and when you must have your ad submitted. They have to have all the ads ready well before the print date, so deadlines are critical. You can't miss them. The good news is that if you can get your ad in early, your ad rep will often be willing to give you feedback and tips on how to improve the ad. This is invaluable information when you're first starting out.

I like to start by writing the copy for the ad. Just like any digital ad, you want to use a hook or interesting headline to capture people's attention. Then, if there's space, include the problem you're solving and the benefits people will get. It's a good idea to reread the "Psychology of Ads" chapter of this book before you start making print ads, just to refresh your memory on emotional hooks.

Finally, for the love of all that's holy, use an obvious call to action! You'd be amazed how many ads never tell the reader what to do. *Call now, call immediately, pick up the phone and dial this number before it's too late.* You have to tell them exactly what to do. It's even better if you can tell them two or three times. But at least once is mandatory.

Now, if you're working on a classified ad, there's no layout necessary. You just write your copy, add your tracking number, make sure it has a call to action, and you're done. But if you're working on any other kind of display ad, you're most likely going to need a designer to help you, at least for your first few ads. You don't have to hire anyone full time, just find someone on Upwork or Fiverr or even Craigslist who can help you do the design and layout.

You want someone with experience working on print ads. Many web designers don't understand image sizes for print—that being the difference between low resolution and high resolution (72 dpi versus 300 dpi). And they may only work in RGB colors (for computer screens) rather than CMYK (for print). Print is just a whole different ball game from digital, so check their portfolio to see that they've worked with magazines or newspapers. Once you find someone you like, you can simply forward the specs and your copy, and let them go to town.

When you get your final files, make sure they give you the actual source files, usually an Adobe Illustrator or Photoshop file. Without the source files, you or your designer can't make any edits, and that can be a problem later on.

Finalize placement

The last step is to pay your bill and get your flight date, which is the date your ad will run. It's important to keep track of deadlines and submit your ads early. That way you can fix anything that's missing or incorrect. If you're right at the deadline, they'll run your ad with the mistakes included.

Once the ad is in print, you're going to review the results at 30, 60, and 90 days. If you got good results, you could scale the ads to bigger budgets and placements. If the results weren't so great, rethink the ad and tweak it for another small test. Just like with digital advertising, testing is key. It just takes longer to get results with print.

Tips for Print Ads

1. If most of the publication is black and white, say for a newspaper, ask if there's an option to print in color. Even just one extra color on a tiny placement will make your ad stand out from everything else on the page.

2. Make sure you have the option to change your ads every print cycle. Some publications don't want the extra work, so you have to run the same ad for your entire contract. If that's the case, then there's no way to test and improve results.

3. Ask your ad rep what the most successful advertisers are doing. That's how you learn this game, by being curious. Which placements are they buying? How long are their contracts? Which works better for a quarter-page ad, the upper right corner or upper left? They may not want to share that information, but some may be willing to bend over backwards to share with you. It doesn't hurt to ask. And most people never ask, so they never know.

4. If you can, find out what content (called "editorial") will be placed near your ad. If you're selling auto insurance, it would be great to have your ad next to an article about cars. There's no harm in asking if your ad can be placed in or around certain relevant articles. Some ads actually read like full articles, but they're labeled "paid advertisement" in small print at the top of the page. These are called advertorials, and they can be very effective, though they are more expensive.

5. Make sure you read and understand the print specs, even if you use a designer to create your ads. You need to be able to speak the language when you're talking to ad reps and negotiating rates. If you don't understand something, ask your designer or rep to explain it to you.

6. Get a proof of your ad so you know what it looks like before it goes to print. So many people skip this step and then find that mistakes were published.

There are so many different kinds of print materials you can place ads into, not just newspapers and magazines. There are trade journals, hobby zines, industry and association newsletters, directories, and local papers that are nothing but ads. If you're willing to put in a little extra legwork up front and wait a little longer for your results, print can be a goldmine for you. Because so few people are even trying to run print campaigns, you have a wide open playing field.

Direct mail

A lot of career marketers I know have this weird obsession with direct mail. And I'm no different. Some folks call it junk mail, but to me it's art. Have you ever really looked at the unsolicited mail you get every day? The colors, the styles of envelopes, the lift-letter devices that they use to get you to actually open the thing. It's just incredible to me that advertisers can use such a simple medium like paper and ink to get you to pay attention to something you don't think you need or want.

Watching infomercials is a favorite pastime, and I frequently end up buying things just so I can get their follow-up mail and see how they're crafting their advertising. I love to read the sales letters and see how they make messy mailers and bulky mailers—it's endlessly fascinating. It's like getting a never-ending master's degree in advertising that comes delivered for free every day. You can learn a lot from direct mail if you pay attention.

Most people don't realize how much volume direct mail actually drives for pay per call. We're currently tracking nine figures annually. It's generally targeted to an older demographic, people who open mail and actually make phone calls or for things that are an immediate need. For example, direct mail is huge in Medicare and drives an immense amount of Medicare traffic. And it results in really high-quality consumer engagement because recipients are literally holding the letter and thinking *I need this right now. I'm gonna pick up the phone.* Offers like auto insurance, financial services, mortgages, and home services all do well with direct mail because the higher payouts balance the production cost.

The downside is that printed mail is expensive to send because you have to pay postage so that people in tiny trucks can drive around and shove them into people's mailboxes. To be successful at any reasonable scale, you'll need to work with a mailing house. (Either that, or trick your kids into licking 50,000 envelopes.)

The process is pretty simple. You send a list of names and addresses to the mail house and they will do the printing and sending for you. They will print all kinds of advertising from postcards to full-out catalogs, and they can use special devices like colored stamps and ink to get the piece to stand out. Make sure you triple check that your data is legit and you're sending to actual households because once you send that mail merge file, you can't take it back.

Six words of warning: CHECK YOUR PHONE NUMBER EVERY TIME! A good friend of mine called me in a total panic because there was a typo on a million mailers that were about to go out. The phone number did not, in fact, connect to his buyer. Instead, it connected to a sex line. A million seniors calling for health care would be greeted instead by "Hey baby, wanna f**k?"

My friend was beside himself. It was a really terrible situation. I had to track down the company responsible for that number on LinkedIn and beg and plead for the leadership of a massive telecom network to please, please, pretty please call the owner of the sex line and ask if my friend could rent the number for three months until the mailers were forgotten.

Thankfully, they agreed, but it cost my friend more than $10,000. So, for the love of all that's holy, proof your numbers! Call them. Call the number on your mailer every single time.

Timing is important with direct mail, too. Notice what you get in your mailbox. Say it's spring in Michigan, and everything thawed this weekend. We're gonna see a ton of direct mail for things like gutter repair, window replacement, driveway resurfacing, pest control, mosquito repellent, and so on. And those perform really well for home services campaigns. Because people need it, right? They're like, *Oh, yeah! My gutters are clogged. I need to call the roof guy.*

Affiliates generally shy away from the unknown. They think everything is digital, Facebook ads or Google ads, but in reality, there's endless millions of dollars to be made in alternative print mediums and direct mail. Print is by far the oldest direct-response marketing channel. It's been going on for hundreds of years, and it's not going to stop anytime soon.

Now, the mailbox isn't the only way to get your message to people offline. The next three chapters will open your eyes to a whole new world that you probably dismissed as too old school to bother with. We're going to start with out-of-home advertising, which is a multibillion dollar industry.

Ready to learn about the power of billboards, benches, and bus wraps? Keep reading ...

CHAPTER ELEVEN

OUT-OF-HOME ADVERTISING

I don't know about where you live, but in my neighborhood, there are all these little yard signs that people put at intersections for plumbing, painting, fixing your furnace—lots of home services specifically. And they all have the same local phone numbers on them. One day, I got curious, called the number, and asked the person who answered if they were the actual plumbing company. They said no, they would answer my questions and set up the appointment. So, right away I knew this was a lead-generation call center. I made some excuse and hung up so as not to waste their time.

Then a little while later, I'm driving behind this guy in an old rusted-out pickup truck, and he's stopping at every intersection, hopping out, and slamming these little road signs into the ground. I had to know more, right? So I pull over and ask him what he's doing. At first, I thought he was the lead generator. But no. He says, "Well I saw this ad on Craigslist. They wanted to pay someone to put up signs. All I have to do is follow this map and put a sign on every intersection. They pay me $200 for the day, and if I do a good job, they send me a few jobs every month."

Then I got curious about the math and asked him how many signs he put up in a day, and he said about 200. So, whatever company is running this campaign is paying a dollar a sign for labor, maybe $2 for the printing, and they're getting leads from all over town. So, if they got a single phone call off a single sign, they're multiplying their money per sign by, like, tenfold.

That got me thinking that nobody is really doing a great job of this at scale, yet they must be working because I see the same crappy signs all over town all the time, and they stay in place for weeks. Always the same design. Always the same call. Of course, I immediately thought of all the ways I could do this 100 times better using QR codes for calls, using different tracking numbers to figure out which intersections work best, and paying people all over the country to do the same thing. I could imagine dominating the crappy road sign industry overnight! (I know. I need a life. What can I say? I love advertising.)

I later learned that that sign strategy is one piece of an entire advertising industry called out-of-home, or OOH. Out-of-home advertising is any type of advertising that reaches people when they're outside of their house. Sounds obvious, right? The fact that we all have phones in our pockets all the time technically means *any* kind of advertising could be considered out-of-home. But since I don't control that industry, we'll just go with their vocabulary, OK?

When you hear out-of-home, think old school. We're talking billboards (both traditional and digital), ads on buses, subways, and other public transportation, flyers stapled to utility poles, little signs stuck in the ground at stoplights—things like that.

OOH also includes gas pump ads, in-flight advertising, airport, arena, and stadium advertising, bathroom and urinal advertising. There's no end to the places you can put an ad. For this discussion, though, we're talking mainly about billboards. Why? Because they have the largest reach of all OOH methods.

You might be thinking I'm living under a rock advocating a strategy like billboards and park benches, but check this out: According to the Out Of Home Advertising Association of America

- OOH revenue increased 20.7% in 2022 compared to 2021.
- 2022 revenue generated around $8.6 billion, a new revenue high that wasn't predicted to happen until 2025.
- OOH media was the fastest-growing media channel in 2022. Its 7% growth outperformed even search engines and social channels.

- The top 10 spenders in OOH are Apple, McDonald's, Amazon, Panera, Google, American Express, T-Mobile, Disney, Coca-Cola, and Universal Pictures. In fact, in 2022, the top spender, Apple, increased its spending 27% over 2021.[1]

And according to 2022 research by OneScreen.ai, 96% of marketers say they achieved ROI goals with OOH advertising, including monthly revenue increases of 50%.[2]

That should tell you something. Billboards and posters are definitely not old-school. They work, or the largest companies in the world wouldn't be using them. Another interesting stat is that 26% of the top 100 out-of-home advertisers are technology or direct-to-consumer brands. Which means OOH is great for pay per callers because you're always driving consumer traffic.

Why is OOH so effective? More than 50% of consumers say they pay attention to digital OOH advertising. If you've ever visited or seen pictures of Times Square in New York City, you're seeing this concept in action. Digital billboards plaster the entire block in every direction. You cannot look in any direction without being hit by at least one ad, more often 10 or 20. They're flashy. They're bright. They show images of sexy people doing sexy things. There's an atmosphere of excitement about the place, especially at night when people are out and about enjoying themselves.

But fancy tourist destinations aside, Americans spend an incredible amount of time just getting to and from work each day. Since 1980, the average commute time has increased 27%. The average one-way commute is 27.6 minutes. The longest, in New York City, is 34.7 minutes.[3] That's a lot of time to be stuck in traffic or riding public transportation. And while they do, they're taking in advertising.

OOH product advertisers see an almost 600% return on investment. Telecommunications ads are seeing a 472% return. Retail companies are seeing 379%.[4] That may not be immediate. These placements get in people's heads over time, and they work better as people see them over and over and over again. You may not see that type of ROI upfront, but over a 12-month contract or a 24-month contract, it may average out to something like this. When done properly, OOH can yield spectacular returns compared to online advertising, but it is higher risk, and it requires a hell of a lot more work.

Now before you get all worried that you can't possibly compete with the likes of Apple and Google, it's important to remember the distinction between branding and direct response. Pay attention to the billboards you see everywhere, and you'll notice that most of them are either raising and reinforcing brand awareness, or they're *completely* ineffective because they're too cluttered, too vague, or lack any sort of a call to action.

Pay per callers are more intelligent than most people designing billboards. They know how to work with direct response, and they can apply the same direct response tactics to the billboard medium. If you don't know yet, you soon will—just keep reading.

Before we dive into the specifics, there are a few things you need to know about the industry in the U.S. It's heavily regulated by individual states. In fact, you can't use billboards in public spaces at all in Maine, Vermont, Hawaii, or Alaska. That doesn't mean you can't use OOH, you just have to be a little more creative in those states.

You can't construct a billboard and slap up an ad just anywhere. They require permitting and licenses, so you have to go through a consolidator or advertising company that actually owns the licensed placements. That's actually a good thing because it takes all the headache out of the process. You don't have to deal with the local government. You don't even have to climb a ladder. You just pay a fee and design your ad.

OK ... maybe it's not quite that simple. But one of the reasons I love OOH is that it does take a bit of extra work to get a winning campaign. It takes extra digging to figure out how this medium works. (It's not like there are a ton of How To Get Rich With Billboards courses or YouTube channels out there.) And as you know by now, I love the channels that require work because it keeps most people out. The fewer competitors, the better.

Now thankfully, most of out-of-home advertising is already consolidated by companies like Lamar, Clear Channel, OUTFRONT Media,

JCDecaux, and iHeartMedia. And so you really only need to know these five companies if you really want to get started in the space. The downside is that it's a very corporate space. You're going to have to deal with corporate sales cycles, and it won't be as easy to negotiate rates. They're used to dealing with Apple and Google, so you're kinda small potatoes to them. But that's OK, you can still do really well in the space if you're persistent and creative.

Pros and cons of OOH

Just like every advertising channel I'm talking about in this book, there are pros and cons. One of the biggest issues, which could be a pro or a con depending on how you look at it, is there's an endless number of placements, locations, and options to choose from. You have a ton of variety, but you also have to sift through it all to figure out your best plan of action.

But think about this: You can hyper-target placements for *exactly* where your target audience is going to be *at the precise moment* they need to make the call. If you're running a car insurance campaign, you can literally buy placements across the street from car dealerships. Running a senior living campaign? You can buy placements outside hearing aid stores, hospitals, or in neighborhoods with an aging population.

Once you figure out a placement that works, you can buy that billboard for an extended period of time. That gives you a monopoly on that placement for as long as you want it. No competitor can come in and just take over your billboard. You have to willingly surrender it. And there's an extremely low chance that competitors will copy you because it's extremely unlikely they will even see the campaign. The likelihood that another pay per call marketer will see your billboard and then do the research to duplicate that campaign is essentially zero. You own the space. Win!

One of the big negatives about out-of-home advertising is that you must have a unique phone number and artwork for every placement in order to track correctly. If you're advertising on a thousand different benches across Chicago, you're going to need a thousand different phone numbers. Otherwise you'll never know which benches are bringing in the calls and which ones you can let go of. You're also going to need a thousand different creatives, so each bench ad has the correct phone number.

If you're advertising in different formats like billboards, benches, and buses, you're going to need different sized artwork, too. The good news here is a virtual assistant can solve that problem pretty easily. Once you have your ad designed, you can simply hire someone to change the phone number on each one.

Another drawback is that your contract length is going to be in direct relation to your placement price. The longer your contract, the less you pay per month. That also applies to multiple placements with the same company. There's a bit more risk here because you can't do quick testing and then run with whatever works. You have to know what you're doing up front. So limit your early ads to one location until you figure out what works, and then think about scaling. (At least you don't have to worry about the timing variable. Billboards work 24/7.)

Testing is also more expensive because every time you redo your billboard, you have the upfront cost for printing and hanging. If you're trying 10 different creatives on static billboards, someone has to climb that ladder to pull down the old ad and put the new one up. That's why it's more cost effective to have longer contracts of 3, 12, or even 36 months. But you better know that what you're doing will work before you commit.

Think about what the consumer will be doing when they see your ad. That drives a number of elements in your creative. If you're working on a highway billboard, people will be driving past it at 70 miles an hour. You can't put a lot of words on there, and your phone number has to be memorable. If they're driving along and see "Call us at (833) 431-2687," they'll be past the ad before they finish reading it. They'll never remember the number unless they drive past the same spot every day and they're highly motivated.

On the other hand, if they see "Need Insurance? 1-800-Call ACE," they'll remember it because the only word they really have to recall is ACE. 1-800 is automatically recorded in the brain because we see it all the time. It's a phone number, so the word "call" is automatically recorded as well.

The next time they think *I need insurance*, it will likely trigger that memory.

But there are lots of OOH placements where people aren't driving and can take the time to look at your ad and take a picture of your phone number or scan a QR code. If you're working with a bench or subway station poster, the phone number is a little less important. Capturing attention and getting your message through is more important.

As you can see, out-of-home advertising requires a lot of thought, a lot of planning, and a lot of talking to people. It takes strategic thinking about how you want to approach each placement in each campaign because they're longer term and it requires actual labor to put these things in motion. And for people like me who love a challenge and hate competition, this is a great advertising medium to get into. Just know it's going to take time to pay off.

Here's the process for billboard advertising

Step 1: Research and choose a location. If you're going to get into this space, I highly recommend that you choose one metropolitan area to start with. Preferably, one near where you live so you can actually go out and view all the placements and watch how people interact with them. The best thing to do is drive around and actually look at the ads, the placements, the traffic patterns, everything. Also look at the other billboards and OOH ads in the immediate vicinity. What's the competition like? What do these other billboards tell you about the demographics of the area? (Because you know those other companies have already studied these stats.) On-the-ground research will give you a better understanding of how to implement an OOH campaign.

When researching an area, think about where people are, what they're doing, and what they're thinking about. If you're targeting older people, you might want an ad near the social security office. If you're working with automotive offers, maybe you want to be near the DMV or auto dealerships. Supplemental health insurance? Maybe hospital and pharmacy locations would be a good idea. Home services? Consider door hangers. It's actually fun when you get into all the little details.

Step 2: Contact all the media vendors in the area. Ask for placement maps and rate cards. Every vendor will have a map showing all their placements and what each one costs, along with different contract durations. These are street rates, and you never want to pay the street rate. But it gives you an upper limit to go by when making decisions. It also gives you ammunition when you start negotiating the rates you'll actually pay. Make sure to also ask for traffic statistics. Just like with online advertising, you want to know how many impressions your ad is likely to get. When you're first starting out, this information won't mean much. But once you know traffic and viewership patterns that work, you can duplicate your success by seeking out similar patterns elsewhere.

Step 3: Assign phone numbers to each placement. You need unique tracking numbers for every placement so you know what's working and what's not. However, if your billboards are across the street from each other or in the same vicinity, then you want the numbers to be the same so people remember them. Also, think about whether or not you need a special vanity number or just a regular number based on the placement.

Every number must be tracked or you won't have any idea how many phone calls you're getting. Whatever you do, don't let a network give you the phone numbers. If they can give them to you, they can take them away. And the last thing you want is to have a three-year contract on a billboard and then suddenly the phone number doesn't belong to you anymore. For OOH ads, there really is no other option but to have your own call-tracking service. Otherwise, you're putting yourself in a situation where someone else can cripple your business and you have no recourse.

Step 4: Create the ads. Billboards are complicated if you've never designed one before. When you work with a national agency or consolidator, they're only too happy to refer you to people who can take care of all that for you. They'll do the design and the printing—all you have to do is ask. But remember when I said most billboards suck? That's because they're created by designers with no training in direct response. Worse, the people approving the ads have no clue either. And these agencies cost a fortune.

My preference is always to create the ad myself with a less expensive designer, and then check with the OOH agency to see if it's within the right parameters. When choosing a designer, make sure they are familiar with print advertising! This is critical. As I mentioned earlier, digital designers work in RGB colors and low-resolution images. Print designers work in CMYK colors and high-resolution images. They are two different animals, and you don't want to spend money only to end up with the wrong files.

Get two people to look at the ad, check the spelling and the phone number, and double check the specs against the ones you'll be given, and then give any and all corrections to the designer. As long as the specs match, you'll be fine. But if it goes to print with a mistake, you're screwed.

One way to save money, especially in the early testing phase, is to use digital billboards. There's no printing involved. And they're similar enough to online ads that most marketers feel comfortable working in this medium.

One of the biggest problems with most billboards is they're confusing. They have too many words, clever concepts, and the call to action is either too small or nonexistent. This is where you'll shine, because you understand direct response advertising. One simple, direct message is what you want—nothing clever or hard to grasp in a fraction of a second. You don't want three sentences, you want six words and a nice big phone number. People need to be able to glance at the ad, understand it, and decide if they're interested. If they have to read it, it's probably not going to work well.

Step 5: Review the results. Once your campaign is running, you want to review the results on a weekly and monthly basis. Hopefully, you'll notice over time that the results increase like a snowball because of the frequency people are seeing the ads. If they commute the same way or walk the same route every day, they become familiar with the ad, and some level of trust sinks in. They may walk by a divorce attorney ad for months and never think twice about it. Then one day, bam! They actually need a divorce attorney, and they'll see the ad at the exact right time. So the longer the placements sit, the more effective they should be. That's why you don't really want to rent a billboard for a one-month period, even if you're just testing. The nature of the OOH industry is that you have to be patient.

Step 6: Negotiate your rates. When your renewals come up, you just look at the data and renew the placements that work. It's that simple. If you have placements that are breaking even after three to six months, hang on to them. They probably just need a little more time. See if you can negotiate a slightly lower rate, and definitely listen to the calls themselves so you know if there's a problem on that end.

If you're buying hundreds of placements and some of them just aren't working at all, you may be able to negotiate different placements or see if the company will let you out of the contract. Of course, the more you buy, the more willing they are to work with you and give you special deals. If they won't budge on rates, see if they'll build you a package deal. You can negotiate for all sorts of things like multiple placements, longer-term contracts, mixed media, and mixed placements. Everything is up for discussion if you just ask. If you don't ask, you don't get.

One superpower direct response marketers have is they're able to take what most people think is a bad placement or media type and turn it into gold. This is really true in the billboard space because your goal is dramatically different from what most OOH advertisers want. Most billboards are branding campaigns. The only goal is top-of-mind awareness and pushing forward a certain image or feel.

We couldn't care less about that, we want people to pick up the phone and call! That means we don't need the most expensive cities like San Francisco or New York. And we don't need the most glamorous placements like Times Square. We need to find the right people when they are most in need and are most willing to make that call.

Remnant inventory can be a gold mine for us. These are placements that are so undesirable that they just don't get bought up. But depending on what we're selling, some of these remnants might be perfect for our needs. Drive by the placement location or hop on Google Maps and look at the street view. What other businesses are around there? What are the general demographics of the people who live there? What type of restaurants are nearby? Who eats there—younger people or older people? A billboard that's next to a neighborhood restaurant that older people visit might not be great for an Apple ad, but it could be perfect for your medical supplemental insurance ad.

Another thing you want to think about is consecutive placement. That means how many times might someone see your ads as they go about their business. If there's a popular bus route that a lot of your target customers take to work, you may want to have an ad in the bus shelter, another one on the side of the bus, and a third inside the bus. Oh, and if people typically transfer to a different bus to finish the route, put one in there as well. Repetition creates memory.

For over 70 years there's been a famous series of billboards on Interstate 95 in North Carolina and South Carolina. They are advertising a rest stop called South of the Border, which is a Mexican-themed tourist extravaganza right on the border of those two states. It's almost impossible to describe this place. It started out as a little shack where weary travelers could get beer and snacks, but now it's an entire complex with hotels and restaurants, a petting zoo, and amusement rides. There's a barber shop, cocktail lounge, post office, and pharmacy. It has its own landmark, too, a giant sombrero sitting on top of a huge tower. Like, you just have to see it to believe it!

The billboards feature its spokes-dude named Pedro, who is always wearing his famous sombrero, and they always include how many miles to go until you get there. They appeal to both parents and kids because the traffic they're advertising to includes tons of families traveling south to Florida and Disney World.

They usually have some sort of cheesy pun like "You never sausage a place. (You're always a wiener at Pedro's!)" And there's a picture of a sausage, you know, in case you're hungry. Or "Pedro's Weather Report: Chili today—Hot tamale!"

Some of them are so direct, it's scary—like, "Keep yelling kids. (They'll stop!)" People actually get nostalgic about these billboards because they've been around since 1949. They saw the signs as kids, and now they want to share the crazy South of the Border experience with their kids and grandkids.

The billboards for this place stretch on for *hundreds of miles* in both directions, northbound and southbound. They're amusing and drive your curiosity if you've never been to the place before. You keep seeing one billboard after the next—175 of them, to be exact—every few miles stretching from the Virginia/North Carolina border all the way to the South Carolina/Georgia state line. At one point there were 250 billboards stretching from Pennsylvania to Florida. Talk about sequential advertising and longevity. South of the Border is *committed* to its billboards!

There are *so* many ways to do out-of-home advertising, we barely scratched the surface. But I hope you can see the potential with this model. It's probably not where you want to begin your pay per call journey, but definitely consider it when you have the luxury of a 12-month turnaround on your investments.

Next, we're turning the dial to WMNY.FM—The Money.

Ready to hop into radio advertising? Let's go!

CHAPTER TWELVE

RADIO ADVERTISING

What pops into your head if I say *I don't wanna grow up, I'm a* ...

Toys R Us kid, right? You're probably singing the song in your head right now if you're over a certain age.

How about *I'm lovin' it* ...

Obviously, McDonald's.

Radio advertising is all about getting a brand, or in our case a phone number, stuck in people's heads—and then exposing people to the same jingle repeatedly. That memorable quality is even more important than the targeting, which makes radio and audio-only advertising unique.

The opportunity in radio advertising is incredible. There are more than 15,000 radio stations in the United States alone.[1] Some of these stations cover entire major metropolitan areas or even several metropolitan areas, while others are hyper-targeted local stations. They all carry different syndicated shows all across the country. Of course, there are international radio stations as well. But for our purposes here, we're just going to look at the U.S.

Now, radio doesn't just mean a broadcast signal flying over the airwaves anymore. You've got three main options when it comes to radio advertising. You have broadcast, which is a signal broadcast over the air waves, like the normal radio you might listen to in your car. Digital radio, which is a signal broadcast over the internet like Spotify, Pandora, SoundCloud, and so on. And there's satellite radio like Sirius and XM radio, where the signal comes from a satellite. (Satellite radio is really interesting because people pay a subscription for it, so they're extremely loyal listeners.)

Just look at some of these radio statistics from MusicalPursuits.com. Sixty-one percent of the U.S. population listens to online radio, which again is just a signal that's broadcast over the internet instead of through the air. Globally, the online radio market is expected to grow 10.5% between 2023 and 2029.

There are some major players in digital radio. Spotify owned more than 27% of the market in 2019. Other major players include Apple, Amazon, Pandora, and iHeart Radio. So, we're not talking about a backwards industry that's better off left in the 20th century. Radio isn't going away, and it is keeping up with the times. Check out some of these stats:

1. More Americans listen to the radio than use Facebook each week.
2. Adults listen to 104 minutes of radio per day, 12.2 hours per week.
3. Radio holds the highest share of collective trust across all advertising channels.
4. 77% of listeners would try a brand or product endorsed by their favorite radio personality.
5. Global radio ad spend is a $36.1 billion-a-year market.
6. 47% of listeners believe radio ads are a fair trade for their listening time.

According to 2021 research from Nielsen Media, radio reaches 88% of Americans each week. That means radio reach is the highest of any medium, including social media.[2]

Have I got your attention yet?

How about this ...

Most people think radio is only for old folks and that younger generations will never start listening to the radio. But the research shows that's completely wrong. Edison Research showed that AM/FM radio reach among Gen Z is 55%, which is really high considering they are the first generation of true digital natives. Even more surprising is that

89% of their listening is through regular radio receivers, and only 11% is streaming.

According to Audacy, an audio advertising research firm, audio channels have the highest level of trust (69%) across all advertising channels.[3] Those trust statistics are huge because most Americans distrust online ads. They find these ads annoying and try to block them as often as possible. But for whatever reason, radio is different. People are willing to listen to radio ads because they understand that ads keep the stations afloat. (The same is true for every other medium, but for some reason the logic doesn't translate.)

The good news is that because radio is more than 100 years old, it's not the shiny new thing on the block. Only the savviest marketers are seeing the incredible opportunity that radio offers. The top four radio advertisers are T-Mobile, Comcast, Home Depot, and AT&T. Three out of four of those have pay per call ads, which just proves that the radio format works with pay per call. It's truly like a wide open playing field if you're willing to take the time to learn the ropes.

It's interesting that even with all this opportunity and me teaching this concept to pay per callers for years, I have never heard a pay per call ad on the radio in my local area. Makes you think.

Pros and cons of radio advertising

The biggest pro is radio's massive reach. Any channel that can reach 88% of the population is a place you want to be! There's almost unlimited opportunity to scale. Most marketers are looking to make a few hundred dollars a day. With pay per call, I was looking for opportunities that could give me seven figures a month. And radio was absolutely one of those opportunities. Once you find something that works, you can just pour more gasoline on the fire. When you do that, the ROI goes way up because the price per ad decreases the more you buy.

There's almost no competition from digital marketers because most are not doing call campaigns, and they're not really thinking about radio at all. Radio is usually considered branding, so people don't associate it with direct response. And there's a bit of work that has to be done to set it up including split testing ads, negotiating with a human (God forbid!), and you may have to put up more money up front. All of that means it takes more time to get results than if you just slapped up a Facebook ad and let it ride. And that high barrier to entry is going to keep most people away. And you know I love that.

Even if another marketer is paying attention—which they're not—it's hard for them to figure out what you're doing to copy you. Lazy marketers love the internet because they can just find the best performing ads and rip them off. That really doesn't happen in radio because you have to be physically located in the area the campaign is happening and be listening at the exact right time to even hear the campaign in the first place. You can dominate major metropolitan markets for specific verticals without anyone even noticing. That, to me, is really cool!

Radio campaigns are very easy to take international. Overseas radio has its own rules, and you'll need a call center that speaks the correct language. But, in general, the opportunities are pretty similar. Even better, the costs are lower and the placements are better. We're not going to talk a lot about international ads here, but just be aware that it's an option.

Here's something that's a little bit scary: With radio, you can subconsciously program people. You can get in their heads and carve out a memory that you own in their brains. People trust what they hear, especially if it's in a voice they listen to all the time. Radio hosts become trusted advisors. So much that nine out of 10 listeners know personal details about their favorite host and follow them on social media.

Not everything is perfect in the radio world, although I think it's pretty close. Here are the downsides to consider when advertising in this medium.

Targeting is not an exact science. We'll talk about this more in a little bit, but just know it's more of an art combined with some statistics. You're never going to dial it in perfectly, so you're going to have to be creative. I actually think this has an upside because it's going to keep out a lot of potential competitors.

It costs more than online advertising. You can't do a radio ad with 50 bucks. You're going to need a few thousand dollars to get into radio, so it might not be the best option if you're just getting started. When you get to the point where you can afford to lose $2,500 or so to test some things out, then you can get into this space without worrying about whether you're eating ramen for the next month. The good news is as your budget goes up, your cost comes down. So even though starting is expensive, things do even out.

Radio is obviously limited to audio, which means people have to remember who you are and how to reach you. It's a little tricky, but we'll talk about it shortly.

It requires a lot of creativity to win in this game and immediate campaign pausing is just not possible like it is with online advertising. Reps can pause an ad, but it may take a day or two. That means you have to be thorough and make sure there are no errors in your ads.

Radio advertising basics

Step 1: Determine your target audience. Just like everything we do in marketing, it all starts with the audience we're trying to reach. Think about what type of campaign you're running. Are the listeners male or female? How old are they? Do they work odd hours? What's their income level? What other demographics can you find out about them?

Step 2: Decide on geographic targets. Where you run your ads matters a lot. Obviously, you don't want to run ads for a Chicago chiropractor on a radio station that only airs in Kansas. If you're not running a campaign with a specific location, though, you have a little more freedom. You may want to start in a smaller geographic location to test your ads because the prices may be lower. Then, when you have a winner, you can scale the campaign into major metropolitan markets.

One more thing to understand about locations is you need to get into the mind of the consumer. If you're running a campaign selling condos in Florida, you might think the best place to advertise is Florida. But if you've ever spent a winter in the northern tier of states, you know everyone wants to head south that time of year. Older people in Vermont or Michigan might respond nicely in, say, February. So it may be worth testing some creative in different markets during different times of the year.

Step 3: Reach out to radio stations and talk to ad reps. Once we understand who our target audience is and where we want to reach them, then it's time to reach out to actual stations. There are a few ways to find stations. Of course, you can just Google radio stations in Tampa, Florida. But that's not the only way. If you're looking to advertise on internet radio and streaming apps, there are self-serve ad platforms that are as simple as uploading an ad and telling the platform where you want the ad served. You can also find lots of stations at once by searching Google for "online radio station directory."

You should never talk to just one station. Three to five is best, so you can compare them head to head and use that information when you start negotiating. When you call each station, ask for the advertising or sales department. When you get an ad rep, ask them for a rate card with reach and demographic information. They will have a whole package of standard information for you to look over. The rates in those packages are called street rates, like rack rates with print advertising. They just give you an idea for the upper limit of what you can expect to pay. You should never pay the street rate! You're going to negotiate it down. Don't worry, we'll get to that.

Step 4: Choose your spot length and placement. A radio spot is going to be 30, 45, or 60 seconds. It's not necessarily cheaper to have a shorter spot, so you want to be strategic about choosing the length. You also need to decide on what time of day you want your spot aired, that's your placement.

Radio stations have segments of ads that play between the regular programming. Typically those segments are made up of three or four ad spots. It all depends on the radio station. What that means is regardless of whether you choose 15 seconds or 60 seconds, you're going in an ad segment with other ads. They may discount a 30-second spot, but it probably won't be half the price of a 60 second one. It might only be 15% cheaper.

Repetition is the key to successful radio, and there's only so many times you can repeat a phone number in 15 seconds. So you really need to look at the math and see if it's worth it to take the discount. A longer, slightly more expensive spot may convert at a rate that makes the higher price worth it.

Step 5: Negotiate the best price with the advertising rep. As I mentioned earlier, you want to talk to several stations and get all their information. If these radio networks think that you're only talking to them, you won't get good results from the negotiation. But if you're putting them up against one another, they're going to try to win your business. That's the position you want to be in.

Radio reps tend to be hungry, especially on traditional radio that broadcasts over the airwaves. They're typically longtime radio sales veterans that have seen ad revenue decline. You'll be able to push them to lower their rates if they know they're competing against rival stations. Just let them know you're comparison shopping, they can't fault you for that. By telling them you're calling other stations, you're letting them know you're serious and this isn't your first rodeo. But it doesn't have to be negative. You can be casual and build some fun rapport.

Step 6: Create a radio advertisement. Making a radio ad is not expensive or difficult. You don't need a professional studio and celebrity voice actors or anything like that. I've run quite a few successful ads that I produced for a hundred dollars and recorded in my own voice.

When my business partner Harrison used to live with me, we decided one day we were going to figure out how to do radio ads. We found this hilarious kid named Mark on Craigslist. He had a band, and so we figured he would have everything we'd need. He brought a microphone over and we recorded 30 or 40 ads in an afternoon. I don't even remember if we paid him or just bought him lunch. But the point is we made a whole bunch of ads for practically nothing, and they turned out to be really successful. You don't need to hire an expensive agency.

The station will tell you what format to save the files in and how to send the ad to them. Pay attention to deadlines. This isn't like digital ads where you can turn them on and off at will.

Step 7: Confirm your time slots. Here's the thing, if you don't confirm your time slots, they're going to assume you don't care and put you in less advantageous spots. Smart advertisers ask for high-quality placement. So if you don't ask, you won't get the best slots. One other thing to confirm is that you get a *permanent first run* as a term in your contract. That means you're the first in any string of ads being aired. That's important because if someone else has a crappy ad, the listener might change the station before they even hear your ad. You want to be first.

Once the spots have aired, you want to follow up with the ad rep and make sure they actually ran at the correct time and placement. And you also want to ask for recordings, so you can keep a record and hear what it sounded like in real time.

Step 8: Finally, review the results. This part's simple. If it's close to profitable, we optimize. If it's way off, we don't. We cut the spots that don't work, keep the ones that do, and scale. One thing to note is you're going to need to run your ads over and over to get good results. The chances of an ad running once and the right people hearing it and calling are pretty slim. Be prepared for that. The second or third time they hear it, they'll think *oh yeah, I was going to call that place*. Maybe the fifth time they actually call. Repetition gets results in radio.

I recommend running an ad that has a multiple-week or multiple-month flight to really see what it can do for you. If you run a spot for two weeks and don't get a single phone call, you know something's wrong. But if that same two-week spot is producing, it may need more time to really take off.

The best results, in my experience, come from picking a couple target markets and consolidating the budget in a place where you can get the spot repeated a lot. Because your radio ad may not work if you play it twice in a week, but if you play it every single day for three months, you might literally train the market to call you. That's when you start getting massive ROI.

Tips for creating great radio ads
1. Record your own or let the DJ read it
Personally, I love recording radio ads. They're fun and you can make a lot of them for practically nothing. I think it's hilarious! But if you don't want to use your own voice for the ads, you can have one of the station DJs read it for you. There are some pros and cons to this.

On the plus side, people are trained to listen to the DJ's voice, so they're less likely to change the station. One research study showed that 80% of listeners trust and value their favorite personality's opinion. And 77% would try a brand or product recommended by their favorite radio personality.[4] That's why talk show hosts and famous podcasters tend to read the ads that run on their show. It sounds like the personality is endorsing the product, even though they're just reading a script you gave them.

On the minus side, not all DJs are great at reading ads. And there's no guarantee that all your ads will be read by the same person, so you lose some consistency. Poor execution can really hurt a campaign.

2. Consider running public service announcements (PSAs)
These are a little trickier because you have to craft your message in a way that's viewed as beneficial to the public. But if you can do that, the ad spots for PSAs are far cheaper. And the audience feels like the announcement is coming from an authoritative body, which can deliver good results depending on the product or service you're trying to sell.

Here's an example:

> *You want to answer that, don't you? I bet it's just killing you. Seeing the soft glow just inches away. Someone wants to tell you something or ask you something. Oh, come on! Answer it already. Just so we're clear. That wasn't my fault. Next time, ignore your inner voice. Don't text and drive! A message from Florida's trusted choice independent insurance agents.*

Don't text and drive is the message. It's an important reminder for the public. And it's in perfect alignment with an insurance offer. So the PSA makes sense. But there's no call to action. It's 100% branding. So for pay per call, you'll probably need to offer something they can get for free if they call in. And you're going to have to work with your ad rep to make sure it falls within the guidelines of a PSA. It's tricky, but not impossible.

3. Make your phone number easy to remember

Repetition, repetition, repetition! That's the key to successful radio advertising. (Have I said that enough times for you? No? That's OK. I'll probably repeat it again later.) But even repetition won't help if you're just reading the phone number in a regular voice. That's where jingles come in. You know those annoying charming musical bits that get stuck in your head for days? Yeah, you want those! You can get people to write jingles for not a lot of money. The station might even have someone who can help. Human brains are wired to remember stories and songs, so if you can make your ad tell a story and put your phone number to a sing-songy tune, you're golden.

Another great way to make that number easy to remember is with vanity numbers. Lawyers and insurance companies love these. It's a number that spells out a word like 1-800-GET-HELP, or it could be an easy-to-remember number like 555-4545. Vanity numbers are expensive, I have to warn you. Ringba has a limited inventory of them, and they can run up to $10,000 a pop just to rent. But when you're using radio, you have to find ways for people to remember the number.

If you're running a vanity number in multiple geographic markets, don't worry, Ringba's reporting will show you what regions and states calls are coming in from so you can still assess your targeting, even when using one vanity number.

Repetition. Music. Vanity numbers. Repetition. (Did I mention repetition? I think I did. Cool.)

4. Be strategic when choosing your format

Are you going to run ads on a talk radio format? How about a sports network? Are your people hip to public radio? Or do you just want to hit the morning drive time on the classic rock station? Once again, you have to get into the heads of the people you're trying to reach. Your demographic research can help you figure out what kind of music they listen to, whether they listen to conservative or liberal talk shows, or if they're listening while shuttling the kids back and forth from soccer practice. It's not that 60-year-olds can't listen to all kinds of different content—of course they can. It's about where you are most likely to hit a large concentration of those people. It all comes down to your research of the people and the geographic location your ads will air.

TYPES OF MUSIC VS. DEMOGRAPHICS

MUSIC TYPE	DEMOGRAPHICS	NOTES
Adult Contemporary	Women 35 - 44	Top 40 Hits with no rap
Adult Standard	All 55+	"Nostalgia" format. Think Sinatra and older pop music
Classic Rock	Men 45 - 54	Rock in general targets men
Country	Areas of low diversity	Largest genre nationwide
News/Talk Radio	Men 45 - 54	Sports fastest growing
Religious	Moms 25 - 44	Religious music, mostly light rock and uplifting
Soft Contemporary	Women 35 - 54	Ballads and easy listening
Top 40	All 18 - 34	Pop and rap music

Two formats I want to call attention to are talk radio and religious talk radio. Those listeners are L-O-Y-A-L! They are not changing the station for anything. They'll leave the station playing in the background all day long. And they actually want to listen to those ads. So, if you're selling something to moms 25–45, you should be all over those religious channels because they're just going to hear your ad over and over and over again. (Did I mention repetition is the key to successful radio advertising?)

5. Don't forget foreign language stations

Very few marketers even bother running ads in languages other than English. But in the U.S., you can make a fantastic ROI by getting a translator and voice-over artist to create your ads in Spanish and airing them on Spanish-language stations. It's even better if your call center has bilingual capabilities. It's a completely untapped market full of people who need home services, insurance, and dental work, among other things.

6. Think about what people are doing at certain times of the day

Timing is everything (not as much as repetition, but you already know that). Why is morning drive time the most popular spot for the 25–55 age group? Because they're getting ready for work, commuting, stuck in their cars, they're awake, and actively listening to the radio. And if they hear an ad they want to respond to, they might do it right then. Whereas if the station is just playing in the background at their office from 9 a.m. to 5 p.m., they're not as likely to be actively listening.

Home Depot runs a lot of ads in the 3 p.m. to 7 p.m. afternoon drive time because people are thinking about home. Late-night slots for advertising are usually cheapest because not many people are awake and listening. But overnight infomercials can crush it when they're done well. You also have to think, are the call centers open at those times? Do people even want to buy your product this late at night?

7. Ask for discounts based on volume

Once you know you've got a winning ad in a great time slot, it's time to scale. That means committing to run the same ad in the same spot for weeks or months on end. When you're buying that kind of volume, you should be getting a discount. Then ask for the same discounts at competing stations, and let them know you're comparison shopping. When you buy enough volume, you can even ask for free or close to free slots whenever they have unsold time.

The goal with radio is to build an audience that knows who you are and what you're selling. That only happens through ... what? Yeah, you got it. Repetition.

Make believe you have a boss. You'd be amazed how much farther you can get with an ad rep when they think you're just some underling account manager doing their job trying to get the best deal possible. Your boss is a bit of a hard ass and doesn't want to pay more than a certain amount per spot. *Mister, can you help me out?*

Just like with print, it helps to mention that you work with a digital marketing agency, and your boss doesn't think radio even works. You want to prove them wrong. You know if you could get some sort of introductory deal and run a few spots, he'd see that it works and then you can place bigger and better ads. You're doing your best to bring the station business. Reps understand this and will usually collaborate with you.

This is also nice because no matter what decisions you make, it's always someone else's fault. You didn't decide to go with the other station, the boss did. You didn't pull that fourth drive time slot, the boss did.

That way you keep your relationship with the rep intact.

8. Expect to fail—at first

Any time you start out with a new advertising channel, you really should expect to fail and lose money. That's just the nature of the industry. It takes time and testing to finally land on that one winning ad that you can scale. Make sure you set a budget and assume you're going to lose it all. If you can't afford to lose that initial investment, don't start with radio. Start someplace cheaper and easier, like online ads.

Remember when I said radio advertising was something of an art? Well, we've just scratched the surface in learning that art here. So, if you want to dive deep into the nuances of creating radio ads, negotiating with reps, and how to target for pay per call, make sure you head over to the Pay Per Callers forum. You've got one free month to get as much information as you can. So, don't wait.

There's one more old-school gem I want to show you. This one is massive!

So buckle up and keep reading …

CHAPTER THIRTEEN

TELEVISION ADVERTISING

After an exhausting day of working a trade show booth in Las Vegas, Harrison and I were relaxing in one of our favorite hotel suites at the Bellagio that we use to entertain clients. I had my feet up on a giant ottoman and was sinking into the couch doing my best starfish impression. It was the height of the Medicare open enrollment season. I casually flipped through channels to look for my favorite thing on TV, the commercials. On almost every news station there was an endless stream of ads trying to coax senior citizens into picking up their phone and calling to enroll in a Medicare plan. Some were hilarious, others were boring, there were even a few with celebrity cameos like George Foreman.

Harrison asked me if I thought any of these commercials actually resulted in any significant call volume, and I replied that I wasn't sure. I begrudgingly got off the couch and grabbed my laptop and started checking to see if any of the toll-free numbers listed in the commercials were routed through Ringba, and to my surprise there were—all over the place.

I quickly logged into our administration portal and watched as a commercial aired at 7 p.m. on a prime-time news network and one of our numbers exploded with call volume at the very second that the gigantic Bellagio fountains roared to life with a thunderous boom. Literally hundreds of callers in an instant. Then boom, another commercial aired and hundreds more flooded through the phone lines. We sat there and watched in awe as tens of thousands of consumers picked up their phones and called the number in these commercials. I guess we had our answer.

We spent the following next few days digging into the TV advertising industry and calling our various customers who were responsible for the ads to learn more about what they did. It was a truly fascinating moment because I had to reconsider the power of all forms of advertising and how they related to getting consumers to pick up the phone and to call.

Before the internet came along, TV was the best place to advertise anything because of its massive reach. But it was incredibly expensive, and you had to have a professional film studio to produce the commercials. You might think television advertising is completely out of reach for anyone in the affiliate world, but that's just not the case anymore.

Television still has amazing reach, it's just more segmented into hundreds of channels. Almost 100 percent of the U.S. population watches TV on a daily basis. Just among Fox, CBS, ABC, and NBC, television reaches 121 million households, and that accounts for more than 300 million individual daily viewers.[1] It's still the largest medium for advertising available anywhere. There's a ton of inventory available, and it works really well for pay per call. People are used to taking action from a television ad.

There are a ridiculous number of platforms and networks available, but they all fall into three main categories. You can choose from broadcast, which is the old-school over the air waves via an antenna that sits on top of your house option; cable TV, which is run through companies like Comcast or Spectrum, and comes through coaxial cable that goes into a box on your TV; or satellite TV like DirectTV or Dish Network that reaches audiences in the United States and Europe from a satellite in space.

One cool thing to be aware of when it comes to TV advertising is that it's the reverse of online ads. With TV you actually pay *less money* to target a specific geographic area, where with online ads you pay more. If you want to reach a really big national audience all at the same time, like say during the Super Bowl, you're going to pay huge rates. So it's important to get your targeting nailed down tight because it keeps your costs low.

There are some companies out there that do television advertising for you. All you have to do is pay them, and they take care of creating the commercial and getting it on TV. But the likelihood that one of these companies will take you on, especially in the beginning, is really low because they want to work with proven advertisers in proven campaigns. There's just less risk for them if they only work with big brands. By now you know that I prefer to figure it out myself anyway, so that's just fine with me.

Pros and cons of TV advertising

The massive reach is a huge pro for TV. Once you find a way to reach your audience and make it profitable, there's unlimited scaling potential. Then you're just one math equation away from hitting the entire United States.

There's also far less competition compared to online advertising because it's a lot more complicated to create a TV ad, and even harder to split test it. And there's more upfront investment you have to make to even get started. You will have competition with branded advertisers like Coca-Cola and McDonald's, but you won't have to deal with some guy named Steve in his parents' basement ripping off your Facebook campaigns. It's really hard to rip off a TV campaign. It's not for the lazy! There's a high barrier to entry, and I love that.

Another pro to TV advertising is you can subconsciously program people to remember your product, your offer, and even your jingle. It takes time to learn this skill, but it's powerful. I haven't been to a McDonald's in over 10 years, but I still know the song—"Two all-beef patties special sauce lettuce cheese pickles onions on a sesame seed bun." It's permanently embedded in my brain. (Yours too. Admit it.)

I'm not going to sugarcoat this, it is expensive to run TV campaigns, especially if you're doing a lot of testing. I think that's a pro because it keeps people who aren't willing to invest out of the game. There are also a lot of compliance regulations you have to follow, depending on the industry.

It's more difficult to make changes using third-party production methods. That means if you outsource your commercial production, it's really hard to change things. You have to go through a company rep, and they have to change the ad and re-render it. Then you take it to your network ad rep, they give it to the network, and schedule it. There are no quick changes or duplicating ads. There's a long process behind it. And that means you have to be on top of your game, and you have to obsess over things like proofing your ads to make sure there are no mistakes.

It's harder to target the right audience with television. There's specialized research that goes into it. You need to know who's watching certain channels at certain times in certain geographic areas. You really have to know everything about your target audience. And the information is not easily located. Even after you've done all your research, produced a kick-ass commercial, negotiated with the ad reps, and everything is perfect—people can and will skip right through your ads. Or they'll get up and make a snack. Or go to the bathroom. So frustrating. You can't force people to sit through and pay attention to your advertising.

You're going to need $2,500 or as much as $5,000 even to run a test. And so you need to do your research to get it right. And there's a bigger learning curve than for online advertising. And that learning curve is going to be slower because you have to wait for the actual commercial to air on television. And then if you're going to do it correctly, you need your viewers to see it multiple times.

It's like a snowball. You're not going to get your results from one commercial; you need to run multiple commercials. People need to see it 7 to 10 times to really know what your penetration rate is gonna be like in a specific time slot. You want people to see it over and over again until they're finally like, *Ah, OK! I'll pick up the phone.*

Immediate campaign pausing is not possible, so you need to make sure that your tracking is lined up with load balancing configured to your buyers and backup buyers. Otherwise, your commercial goes live, and you get calls, but if your buyer goes down or there's an issue with call routing, you won't get compensated for those calls. And you just lit all that money on fire.

So why do people do it? Because you can make multiple millions of dollars with a well-done TV commercial. Like, we're talking *yacht money*. (Some say you're not successful in life unless your yacht has a yacht.) The reach is just that massive. So, to me, it's worth my time and money to figure this medium out. But it really isn't for beginners, unless maybe daddy was a TV executive and you already know how to play the game.

I just want you to realize there are lots of levels to pay per call, and there are always new challenges to meet and mountains to climb. It never gets boring. And there are boss-level playing fields like TV where you really get to be creative, have fun, and make a lot of money.

Here's an interesting research exercise—watch Fox News. Not for 20 minutes or an hour, for an entire day! This might be painful. But forget about your political views and just pay attention to the ads and the targeting. Fox News has a *very* specific audience that's mostly older white men. And almost every single commercial is a pay per call campaign. If you want to get really great data on this (or any) network you're considering, record a week's worth of programming. Send the recording off to a virtual assistant who does video editing and get them to cut out everything except the commercials. Then you can sit back and watch them. Yes, just the commercials. You'll find out exactly what other advertisers are doing, and you'll gain more insight into the audience for that network than any other method.

How TV advertising works

I'm just going to give you the basics here so you understand the process. Then, if you're really interested in diving into television for pay per call, you'll have a basic understanding under your belt.

Step 1: Contact TV networks or advertising brokers for rates. Now, let's be real for a minute. If you're reading this book, you're probably not spending millions of dollars on TV advertising, so the direct networks may or may not talk to you, or they may not be the right fit. Believe it or not, brokers and aggregators can actually get you better deals because they buy so much advertising. TV is super old-school. If you do not have the volume, no one cares about you. So you need to find brokers and contacts who can help you through that process. Don't just talk to one, talk to many so you can learn as much as you can from them.

Step 2: Determine schedule, show, and distribution area. This is the time to think about your schedule. What shows are you planning to advertise around, and what geographic areas are you thinking about? The less money you pay for advertising, the less control you have over these things. If you want the cheapest spots you can get, you're not going to be able to choose any of these things. You're going to have zero control, but you'll get the airtime at a bargain rate. You're also not going to get a lot of repeat views unless you spend a significant amount. If you can make that work—and some people do— you can make good money. But it's going to be pretty challenging. You need an offer with a very broad appeal so the targeting doesn't matter as much.

Step 3: Comparison shop different networks, mediums, and brokers. Call around and get a few proposals for different tiers. Find out what you can get for $1,000, $5,000, and $10,000. Ask where the price breaks happen. Whether you're ready for that level or not, you should understand it so you can build up to it.

Step 4: Finalize your placements and flight dates. You need to know that in two weeks you're going to run a spot on the game show network at 8 o'clock at night. That spot's going to play three times every day for seven days. It is much more effective to run TV advertising to the same audience multiple times than it is to run the same advertisement to lots more people one time. Repetition is the key to success with this. They need to see your ad over and over again before they'll pick up the phone and call.

Step 5: Create your commercial with tracking numbers. This can get a little tricky. And it can seem daunting to actually make a TV commercial. But before you freak out, let me tell you this. I recently made a TV commercial purely because I was curious. I hadn't made one in a long time. I did it entirely with stock video, so it only cost me about $480. I just taught myself iMovie and figured it out. It turned out pretty decent, too. It's not going to win any awards, and I don't know if it will convert. But the point is that with about $500 and a few hours of time, you can make a commercial. You're going to spend a lot more than four hours on yours, but just don't talk yourself into thinking it's too hard.

Step 6: Deliver your commercial to the network. Once you've finalized, checked and double checked your commercial for errors, it's time to turn it over to the network. Make sure you've talked to them about the format, resolution, and necessary specs to run properly. They're going to bill you the agreed-upon rate whether you hit your length or not. If you're a few seconds under your 45-second limit, they aren't going to say anything. Triple check your work.

Step 7: Confirm your flight dates. Talk to your rep and confirm when your commercial will air. Sometimes networks and brokers will shuffle the schedule and not tell you, so make sure you confirm your dates a couple of days before your flight date. If you don't, they'll just assume you don't care if they mess with your timing a little bit.

Step 8: Ask for viewer statistics and review results. You want to know viewer statistics on the date and time your commercial aired. If they messed with your schedule, you want to know so you can use that as leverage when you negotiate your next slot. Then compare that data with your conversion results. Go into your call-tracking software and see what the call traffic is like. It can take people 24 to 36 hours to call in sometimes. Maybe they're watching a recording of the show a day later. Or maybe they had to wait for a convenient time to call.

Step 9: Determine placement ROI and negotiate new rates. If you're smart and you put a different tracking number in for each placement, then you will know which placement got you the ROI you want. Then you can go back to the network with a bigger budget and negotiate the rates down based on the results. Longer commitments + more spend + lower rates per spot = higher profit margins for you.

OK, let's look at the process a little more closely.

Choosing a market

This isn't complicated, but you do need to put some thought into choosing your market for TV. The four critical questions to ask are: Who are your customers? What do they want? Where do they live? And what do they watch?

I highly recommend you talk to several brokers and ask them these same questions. If they all give you similar answers, then you can be fairly sure the information is accurate. It's in their best interest to help you be successful. Not because they're all that excited about a $5,000 commission from your business. That's peanuts to them. But if you're successful and become a regular advertiser, it could be worth *a lot* more to them.

Do your homework before you talk to any brokers. You want them to take you seriously.

Who's likely to need what you're selling? If you're promoting debt relief, who are those customers? They're different from the ones likely to buy insurance for their boat.

Where do they live? If you're trying to reach baby boomers, Florida and Arizona are good bets for you. If you're trying to reach the wealthiest 1%, San Francisco and New York City are great places to start.

What do these people watch on TV? Older people might be watching the Game Show Network or reruns of their favorite shows. My dad watches reruns of *Star Trek: The Next Generation* all the time. Businesspeople tend to watch CNBC; Republicans watch Fox News. More women than men watch *The View* and Hallmark Channel. Maybe your target is watching fishing shows on Discovery Channel.

This is never going to be an exact science with TV because we don't have the same granular data that we do with online advertising. So just write down your best educated guesses and then go ask some brokers. Get the demographic data from them and ask about the rate differences for major networks versus specialty shows on cable and dish. Fewer people watch those niche shows, so you can reach your demographic for less.

Producing commercials on a budget

We're not in YouTube land anymore. Producing a TV-ready commercial on the low end might cost $2,500, if you find some local kids making music videos and convince them to make your commercial. A professional production for a local car dealership will run upwards of $40,000. National brands will spend half a million or more for big Super Bowl ads, but that's not what we're doing here. Don't let those numbers scare you, because there are a lot of options. And if you're willing to do some work, you don't need a lot of money to do this.

The first thing you need to do is write a script. It takes less effort than you think, if you watch lots of commercials in your niche and pay attention to how they're presented. The average person speaks 150 words per minute. And you're shooting between 30 and 60 seconds—so that is not a lot of words. You have to keep your message direct and to the point. I walk through lots of different styles of commercials and show you what works and what doesn't inside the Pay Per Callers forum, so be sure to go watch those training videos.

You will need a newer iPhone with an advanced camera and something called a gimbal. A gimbal is a gyroscopic stabilizer that keeps all your shots smooth and professional looking. They even come with automatic motion sensors now, so you can set up the camera on a table, and it will follow you around the room. If you're going to shoot with a still camera, then you can get away with just a tripod, but I recommend a motion sensor gimbal because you can use it for all sorts of video content. They're inexpensive, and you can get them on Amazon.

You're also going to want to have a decent lighting kit with two lights on extendable poles and a decent microphone. There are hundreds of videos online about the best way to set up your lighting and sound for filming in different conditions. You may want to check those out before you order anything.

Think about where you want to record your audio. It's really important to have that sound rich and clear. The room you're using should have at least a carpet and some items on the walls to prevent an echo. If you have a co-working space anywhere nearby, you may be able to rent everything you need for the cost of a day pass. Most co-working setups have podcast and video filming capabilities these days. And if you're lucky, they may even be able to point you to a good video editor.

Talking at the camera for 30 or 60 seconds makes for an incredibly awkward commercial, so you're going to want what's called B-roll footage. That's all the extra video that gets spliced into your regular message. Watch any news program or commercial and you'll see lots of ideas for B-roll.

I like to use stock video footage for B-roll and royalty-free music to fill out the main video message. It's inexpensive, and there are lots of websites that sell stock footage and music.

Royalty-free means you pay for it once and you can use it for multiple projects without paying additional royalties to the creator. Make sure you read the standard license agreement when you purchase any royalty-free images, video, or sound files. Different licenses cover different uses, and you may need to purchase a higher level to use that content in a television commercial.

Who's going to deliver the message? This is important. You can be the on-camera personality if you like. You can also hire a spokesperson or actor to read your script. Or you can use stock footage for the entire commercial. It's up to you. You may also want to hire actors to speak your testimonials. You can find people who will do this fairly cheaply on Fiverr. Or hit up your nearest university's film or theater department for some local talent.

Editing your commercial

Once you have all your footage together, it's time to edit. Don't be scared of this part. It's actually really fun once you get the hang of it. If you make a copy of the original footage and use that for editing, then you don't have to worry about messing anything up because you still have the original files.

The most user-friendly editing software I've found for Mac is Final Cut Pro. If you're new to editing and using a PC, then you should probably look into Camtasia as an easy-to-use option. There are just so many video editing programs out there, use the one you're comfortable with.

We're only talking about 30 to 60 seconds of video here, there's just not that much to it. You're going to run a video track, an audio track for the voices, and a music track. Keep the music low so people can hear the message. Splice in some B-roll every 5 to 10 seconds. Type in some text overlay to show the call to action and the phone number, and you're good to go.

The final step is to layer it all together and export it as one file. Follow the station's specs regarding what the final format should be. Homegrown videos actually do really well on TV. You shouldn't feel pressured to hire a big fancy studio or production team, at least not until you have a few tests under your belt. Once you know what you're doing and you're ready to scale, then it may be worthwhile to pay more money to get a commercial that can run for a year or more nationwide.

Negotiating rates

Just like all offline advertising mediums, you can negotiate your rates. The more you buy, the less you pay. And if you can get in with a broker, they can negotiate discounts for you because they simply have much bigger budgets to work with. They will bring a network $100 million a year and get great discounts because of it. You don't have to spend that much, just be part of that broker's list of clients.

Typical ad reps are almost always willing to work with you, too, because they want your business. But with TV, smaller accounts aren't worth much to them, so you have to work a little harder to build a relationship with them. Be honest and tell them your situation, what your budget is, and what you think it will be if these first few tests work out. Never waste their time!

If your budget doesn't make sense for a prime-time commercial, tell them you're willing to take remnant ad space that they can't sell that week. This can work really well if you have a national campaign and you play your time zones right. Do whatever you can to make that rep feel valued. Thank them for their time. Praise them to their boss. And refer people to them if you can. Great relationships pay off big time, especially in advertising.

OTT and streaming platforms

There are other forms of video-based commercials I want to touch on briefly. OTT stands for "over the top" and implies that the video content is going above and beyond the standard TV offerings. We're talking about Hulu, Netflix, Amazon Prime, and other video-streaming sources that skip over traditional broadcasting, cable, and satellite delivery methods.

The cool thing about these types of advertisements is you can target them much tighter than regular television. With television, you're only targeting based on DMA (or designated market area), which is zip codes, and some demographic information based on geographic or subscriber location. And that can be limiting. For example, where I live, we have an older population. We know this because of census data. So anyone buying ads on TV in my town is going to expect the people watching to be 65+. Perfect for Medicare and retirement home ads. But that doesn't include me. I live there. I watch TV there. But I'm not in the overall demographic.

But with TV that's streamed over the internet, you can target much more accurately. You have data on individual IP addresses and how many and what type of devices they have in their home. If they have Hulu on their iPhone or their iPad, you get more metadata about them, which allows you to target even better.

Basically, someone running Hulu ads can say *I'm targeting Adam Young, and yes, he lives in a tiny Michigan city filled with a bunch of geriatrics, but in reality, Adam is 38 years old and likes technology.* And so they can target me specifically based on my actual demographic information instead of inferring it based on the geographic location of the TV distribution.

QR codes to call links work great with OTT because the overall demographic skews younger, and people always have their phones with them. So, as you're running different advertisements inside of Hulu, you can put the phone number on the screen, but also add a QR code to call. When someone scans the QR code, it pops up this little notice on their screen that asks, "Do you want to call this phone number?" That makes it really easy for people to take action.

OTT advertising is relatively new and not a lot of pay per callers are using it yet. Not even all the streaming channels have it available yet. But it's just a matter of time before they do because everyone needs ways to monetize their content. If you're happy being more of a pioneer, you can get into this space early and figure it out long before anyone else.

YouTube

Technically, YouTube is an OTT platform, but it's special because you can retarget people based on targeting pixels. You can target interest groups, retarget them, and create look-a-like audiences. And you can take the interaction rates from YouTube and then retarget with DoubleClick and Google Ads. Everything is integrated into Google Analytics and inside Google Ads.

You can target by channel, by age, and by income level—the options are just freakin' amazing. And so you still need an awesome creative. The video has to be rock solid. You can create it yourself with your phone and some basic editing software, but people have trained themselves to click that "skip ads" button, so you have to grab their attention in the first two seconds.

Once you nail the creative, YouTube then becomes this incredible targeting engine that will find your customers for you. You do have to give it a little bit of time to automatically optimize. But if you give it really, really amazing creatives and good targeting and data and guardrails in your settings, YouTube can go out and find unlimited customers for you.

YouTube hosts a huge number of pay per call ads. I don't think that's ever going to change because most people watching aren't really doing anything. They're just sitting there waiting to be entertained. So, if you can get that emotional reaction out of them, get them excited to pick up the phone and call, they're gonna buy. They just are.

You can think of TV, OTT, and YouTube as the same medium but with different targeting opportunities. Take your time to think about who is going to see that video, what they're watching, and what emotional hook you want to use, and you'll tap into a gold mine.

Like I said in the beginning of this chapter, TV is a massive opportunity for pay per callers. But it's a boss-level game. It's probably not going to be your first campaign or even your fifth. Figure out everything you can about your target audience using less expensive media first. Run your tests there and then expand upward. Television isn't going anywhere, so there's really no hurry to get into it until you're ready. But when you are ready, buckle up! It's a hell of a ride.

So let's take stock for a moment.

You've built a professional brand for yourself so people take you seriously.

You've lined up a network or two, and you're starting to get traction with direct buyers.

You're routing and load balancing your calls inside your own tracking software.

You're doing routine QA checks to make sure everything's working as it should.

And you're exploring all sorts of exciting advertising options.

Congratulations! Everything is going well, and you're ready to scale up big time.

How exactly do you do that?

Keep reading …

CHAPTER FOURTEEN

BUILDING A BUYER NETWORK

Just about every day someone in our Ringba family does something that astounds me. Our pay per callers are so talented and creative, they continually raise the bar on what's possible. So much so, that we created the G.O.A.T. Club, which stands for Greatest Of All Time in the pay per call space. This award is given to businesses that create over a million dollars in a single campaign. At first, I thought it would be an extraordinary thing; but as it turns out, our customers do it pretty often.

Take Matt from Futuredontics for example. He currently runs 1-800 DENTIST, which is a massive success in the dental space. It almost crashed and burned during COVID, when all dentists had to shut down except for emergency procedures. The whole industry was in pieces. But Matt was able to come on board as CEO, build sophisticated IVR systems using Ringba, and use all his knowledge of the pay per call space to build it back up from losing hundreds of thousands of dollars a month to the huge success it is today. Just like everything in this space, it took focused effort. But he applied the same techniques you're learning in this book and turned a dying company into a G.O.A.T. campaign.

Deciding when or even if you want to scale your business is a personal decision. It should be based on your goals, your lifestyle, and your ambitions. There's nothing wrong with building a simple set-and-forget business that mostly runs on autopilot. There's also nothing wrong with dominating vertical after vertical and employing hundreds of people to keep your empire humming. It's all up to you. Don't let other people's opinions of what you "should" be doing affect your decisions. You're the boss. You're in control.

But just for fun, let's assume you want to scale this thing to the moon. Here's how to do it...

Creating a buyer network

In this case, we're not talking about creating your own network full of affiliates. This is a managed network of small buyers. That could be a single dentist's office or a chain of them, an insurance office, real estate office, eight or nine real estate offices, 600 real estate offices, or whatever you like. You combine them all together into a buyer network that you're going to load balance your calls through with technology.

There are a lot of pros to having your own buyer network, but the biggest one of all is that you're going to be direct to the advertiser with no network or broker in the middle taking a cut of your earnings. That means you're going to make the most money per phone call that you possibly can.

If you're running a nationwide campaign of dental calls through an affiliate network, you might be getting $10 per call. But if you own the buyers, you went out and signed them up directly, you might make $40 or $50 for the same exact calls. Depending on the type of dentist, it could go up to $80 or more. There's a huge amount of money in building your own buyer networks.

Now, this is also a massive competitive advantage because you can then become the broker. If you have 50 to 100 of your own buyers in a vertical, it's essentially your own offer. So, you can reach out to pay per call networks and say, "Hey, I have a dental offer. It has coverage in these three states. Can you please send me calls?" Basically, you're sending your own traffic *and* leveraging the network's traffic, too, so you're squeezing the most money out of your efforts.

There are some really big companies that do this. But not a lot of affiliates or smaller companies think this way. And honestly, without Ringba, you don't have the technology to support it. That's a really big benefit to you if you're a Ringba customer. You can build your own buyer network early on because we have permission-based control so that your buyers can decide how they want to receive calls, how many calls they're going to get, concurrency, bidding, all the little details. That means you're not doing a ton of account management.

This is why I sometimes call this model a self-serve buyer network—because the buyers have the ability to pause, unpause, and change settings like pricing on their own.

The last huge advantage to building your own buyer network is that it builds enterprise value, which is the amount of money that you might be paid if a company wants to acquire your business. When you have your own buyer network, you're actually a pretty solid acquisition target because it's a lot of work to build these. It's easier for companies that want to get into your space to just buy you out than to build it themselves. If you're just an affiliate, you have no enterprise value. You're working to send traffic to other people's buyers and the business itself is essentially worth zero.

Now, this may not be feasible from day one, but it's not as hard as it may seem. If you're selling dental calls to a network, there's no reason why you can't just add one independent dental office into the fold and then route your calls automatically. It's fine to wait a little while, but realize that the longer you wait, the more opportunities you're missing.

Now just like anything, there are negatives to building a buyer network. There's a lot of business development you need to do. You're going to have to talk to hundreds of companies to get a lot of coverage. The more coverage you have, the more enterprise value you've built.

If you can get to a place where you have close to 100% coverage in the United States, you're talking about thousands of independent buyers. You have effectively built your own category-dominating empire at that point. The upside is really big, but that's going to take a ton of work. (Fortunately, you're not afraid of work.)

Buyer networks also require more explanation, onboarding, and handholding than just going to a call center and saying, "Hey, I want to sell you calls." Call centers already know what calls are, they know the metrics, they know everything. But if you're calling an independent dentist or a tow truck operator, they've probably never bought a phone call before.

They may not understand the value. And even if they do sign up with you, they may not know how to close a sale on the phone.

So there's more work upfront getting your buyers up to speed, but they're paying you significantly more per call, and you're building loyalty. As long as you keep sending them quality calls, they're never going to leave you. You're building long-term value for yourself.

There's also more work in account management. If you're not using software that's designed for a self-service buyer network, you're going to do a boatload of account management because you have to change people's concurrency, pause them, and fiddle with their settings on a daily basis. Plus, you'd have to communicate with hundreds of buyers. That's why pay per call networks don't really do this unless they use our software because it's just way too much work to manage all those accounts.

Remember, I was an affiliate first. Then I built the software I wanted to use. We are designing the future of our software so you can build your own self-service buyer networks and make those huge margins because that's where the future of this space is. Nobody likes losing money to the middleman.

You're also going to have to do a bit more accounting. Any time you're working with a large number of people, there's more accounting. Software can help you with this. You can use QuickBooks to invoice people and manage your payments. You can use Ringba to track all of the calls, obviously, and to see what's owed and what's not owed, and to do all your billing. But you're going to have to bill people and take their money.

If you're going to build a big network of buyers, we're talking tens of millions of dollars or even hundreds of millions of dollars. With that kind of volume, you're going to have to set up your own accounting process, and it will evolve over time. Every time your business gets 300% bigger, you're going to be redoing your accounting process. You should just expect that. It's not really a con, it's just part of growing a business.

I don't want to sugarcoat the amount of work that goes into building a self-serve buyer network. But when you develop statewide, multistate, regional, or national coverage, and you own your brand and you own the network, you've got real enterprise value. At that point, you've got a business you can go out there and sell for a lot of money. So it's totally worth it. Plus, if you love marketing like I do, it's a lot of fun, too.

How to Build a Self-Serve Buyer Network

It goes without saying, but you've got to have technology capable of handling this kind of network. You could invest millions into developing your own software or use Ringba, but you have to have it one way or another.

Next, decide on a geographic area. You don't want one dentist in Pittsburgh, one in Los Angeles, and one in Boise, Idaho. Consolidate your buyers in one area so you're only driving traffic in one market. You can build out over time, but start where you can control things more easily. Build your brand in that one market, and you can pretty much take it over. In all 300 verticals and thousands of sub-verticals, only a few big companies are doing this well. There's a ton of opportunity here.

Once you've decided on your market, you need to determine call pricing. How much will it cost you to generate a call—either on your own or through a network with affiliates. That's your buy price. Then you figure out your sell price—what are your buyers going to pay you per call? That sell price should take into account competition and location. Dentists in Beverly Hills have more money, and media there will cost more than say, Kansas City.

Once you have 10 or 20 buyers clustered into a smaller area, they will have to bid against each other for those calls. This automatically raises your prices for you. That's the beauty behind clustering your buyers in specific markets: You create your own competition, which drives the price of calls up while your cost remains exactly the same. You just reap the benefits of bigger margins.

The next step is making a list of buyers. Don't make this complicated and don't buy a database. Just open up a spreadsheet and start researching yourself. It just doesn't take that long. You're working this like an old-fashioned door-to-door territory. You take your list of buyers and go visit them. Build relationships and keep calling on those customers until you land them. It's just standard cold calling. You should do it yourself at first so you understand the process. Then write some scripts and train a team to do it for you.

Once you've got your first account, it's time to deliver calls to that buyer. You can take a small prepayment from a buyer, like $200, and drive them a few calls. Ease them into it, especially if the whole idea is new to them. Make sure you listen to every call in the beginning. See what the sales team is doing. Are they closing a good number of calls? If not, they're not going to keep buying from you. But if they sell 80% of the phone calls, and they make a lot of money, they're going to be excited about buying more phone calls from you.

I suggest running a small test with every new buyer. Tell them you'll send them five calls over four days. Then you'll listen to the calls and do a follow-up chat to review the results. Then, when you meet with them after the test, you can tell them exactly what happened. "Sarah was amazing. She closed every single call. James is not. He doesn't understand what's going on, and I have notes for you so you can train him. If Sarah had taken all the calls, you'd have closed all of them and made X dollars. We just need to train your people a little bit more."

Running small tests like this demonstrates exactly how powerful your service is for these businesses. They can see exactly what you're going to do and exactly how you can help them. Who wouldn't want to test out the opportunity for just a couple hundred dollars? You can even do it for free and then there's zero risk to give you a shot.

Once you have a few clients in one area, you just expand the territory. Document what you're doing step by step to create a repeatable process. Once you have it down, it's easy to build a team to do this for you. You could even have your very own call center doing nothing but enrolling companies into your buyer network. It costs some money up front, but over time you could generate tens or even hundreds of millions of dollars.

The final step once you have the whole operation running smoothly? Start all over in another vertical. It's smart to move to an adjacent vertical—keep everything in general health care or home services—because you'll already be familiar with buyer intent and where to buy media.

OK, let's dig deeper into each of these steps.

Deciding on a market

The easiest geographical markets to enter are places you've lived and worked before. You already understand where the customers are located, you know the neighborhoods and overall geography. It's much easier to build a territory when you know the area. Now, if you grew up in downtown San Francisco, that may be a little bit trickier to get started just because of the immense cost of operating a business there. You may not find many plumbers that are actually located in downtown San Francisco just because of the real estate prices. In that case, shift your focus to the surrounding suburban areas.

You should have a couple of options picked out, then look for a high concentration of buyers. Cities are generally going to be better than rural areas. Montana is a great place to grow up, but they may not have a high concentration of buyers for you. In that case, you'll just have to pick another city. You're looking for an area with at least 10 to 20 buyers in it. If it only has two, your business depends on those two buyers—and some people are going to tell you no. That's just part of the business.

If it's a thriving market, you should be able to identify buyer groups, like six dental offices owned by the same person. Those are easier to sell because you only have to convince one person and you get six times the volume for a single buyer. Suburban areas are great for this. Plus, the farther out you go from a city center, the easier it tends to be to convert a customer. There just aren't as many salespeople calling on those businesses, so they're not as jaded as downtown offices can be. My recommendation is to start with medium-sized suburban areas.

Another consideration is the overall income level of the area. If you focus on high-income locations, those businesses you contact will be charging more for their services and will be able to pay more for calls. It will probably cost you more to generate calls for them, but you should have a wider margin as well. Lower income markets have much thinner margins, so it's more difficult to get traction.

Determining call pricing

Determining your call pricing can be a bit tricky. There's a bit of an art to it. The easiest way to figure this out is to do a little undercover shopping with another network in your market. For instance, if you're working with dentists, all you have to do is call any lead-generation offer for dentists as a buyer. Tell them the area you're calling about and ask for information like how much they charge per call and how many calls per week they can deliver. The information they give you is your baseline.

Tell your buyers up front that prices can change. In fact, build that into your agreement with them. They have control over their bids inside their account, so they can always raise or lower their spend based on how many calls they want. As competition changes, you'll adjust your pricing. If you have more buyers than calls, you can raise the prices. If you have more calls than buyers, you can decrease the prices so they're willing to buy more volume. Over time, the buyers become dependent on your calls for the growth of their businesses, so there shouldn't be too much pushback unless another pay per caller is working in your exact market.

Just make sure you set the expectation up front that this is a dynamic industry and prices will never stay static.

Making a list of potential buyers

This process is really simple. Open Google Maps and look at the major metropolitan area you want to be near. Let's say it's Kansas City. Now zoom out a bit and look at all the suburbs that exist just outside of there. The bigger the text, the larger the population. So, outside of Kansas City, there's Independence, Overland Park, Lee's Summit, Riverbend, Liberty—lots of them! Great. The largest suburb seems to be Independence, so we'll go with that to start.

There's no rhyme or reason to this, really. Just pick a place to start. Then open up a spreadsheet and start googling dentist offices or tow truck companies or whatever you're working with. There are like 60 dental offices within six miles of the center of Independence. That's a great territory and an exciting opportunity because all those businesses are fighting over the same customers. They're going to want your services. And this is just Independence. There are 50 more metropolitan areas, which equates to thousands of dentists just in the Kansas City metro area. All of them want calls every day. Can you see the power of this simple process?

Make a complete list of every business that might want to buy your calls. It's not going to be so big that it's unmanageable, but you want at least 50 businesses to call. Make sure to record their business name, address, phone number, website, email, and any other relevant information you can pull off their website.

I also record their star rating and how many reviews they have because that information could be important in determining budget. Maybe Independence Family Dental Care has 153 reviews, they're mostly positive. That means they're able to handle a lot of clients. Aspen Dental only has 11, but they pay for ads. That means they want to grow. Maybe Dr. Jones only has one review and is probably not a very big office. He's not going to want to buy a ton of phone calls. Use the review volume to prioritize which people to call first based on how many calls they may be able to handle. Of course, you're going to load balance all the calls with technology, so a one-person office can still be a buyer. But you want to approach the larger opportunities first.

You could hire someone to do this research for you. But realistically it's only going to take a couple of hours, and you'll get a good feel for the area. I really recommend doing it yourself.

Then it's time to start reaching out and selling your services. Document your process as you go and build a repeatable system so you can build a team in the future. If you're going to get thousands of businesses in your buyer network, you're not going to do that alone, so you might as well start prepping your training materials now.

Contacting prospects

Contacting prospects can feel like a complicated process. If you've never done cold calling before, I think you should do this just to get the experience because most successful entrepreneurs have cold called for something. I was a door-to-door salesman in downtown Detroit. I literally worked 100% commission in the winter selling phone service through bulletproof glass. And it sucked, but it taught me how to be extremely resilient, not worry about getting told no, and to be a just absolute killer salesperson. If you're unsure about doing this yourself or you don't like cold calling, get over it because it'll make you a monster in business.

Here's all you have to do: Call each office on your list and ask for the office manager. If they ask you why, you tell them, "I'm with a local agency. We specialize in dental marketing, and we have an overflow of calls. We work with a lot of your competitors in the area, but we have too many phone calls from people who need a dentist. We'd like to send you folks some."

That's a really weird phone call because of the way I positioned it, which will get the office manager on the phone. I personally address this part differently than a lot of people will because I believe in long-term vision as opposed to short-term gains. I just straight up give them some free phone calls to show them where the value is.

"Adam, you're crazy! Just giving people free calls? I could sell those calls to a network. I can't just give away calls!" Blah blah blah ...

Look, when you're cold calling people who've never heard of you, and you're interrupting whatever they have going on, and they have zero interest in talking to you at all, would you rather try to get them to pull out a credit card or just give them something?

Here's how it works:

"We'd like to give you the first five calls free. They're $50 apiece, so that's a $250 value. We have so many calls that we just need a home for them. I'm not even here to sell. Just take my five free calls, and I'll give you five free clients, OK? No questions asked. Is that OK with you?"

And the office manager's going to say, "Yeah, sure, what do we have to do?"

You say, "Nothing at all. We'll just make your phone ring five times, and I'll call you back in a week to let you know what happened."

And they'll be like, "Oh, OK." A little confused, probably.

And then you'll have a follow up email saying, "Oh hey, Judy. Thanks for talking to me today. Your five calls are going to start being delivered during your business hours tomorrow. I'll follow up next week to see how it went."

I would add the business hours into the spreadsheet so that the research doesn't have to be done twice, and then our team can set up the hours of operation for the client.

Then after they get the calls, you email her again. "Hey, you got five phone calls. Our quality assurance team listened to them. Did you know that your team closed four out of five phone calls? They got four new appointments out of five. Two of them were families. You got eight new customers and, from the sound of it, you guys charge $350 a cleaning. So, I just gave you thousands of dollars' worth of revenue. Can you spare 15 minutes to talk to me about it? Just want to review the results."

She's going to be like, "Hell yes!"

Or you can do it the hard way. You can call them up and say, "Hey, I'm Adam from Media Busters, and we specialize in dentists. I got a ton of phone calls. I have so many that I don't have dentists to buy them so we're selling them at a discount right now. They're usually $50. I can sell them to you for $40 a call. There's a $250 minimum. Grab your credit card for me, this takes a few minutes to set up."

You're gonna hear a lot of "Not interested" … click.

Invest in your future, and give away some free calls. It's going to open up every door. No dentist office or plumber or kitchen remodeler is going to turn down free business, especially if they don't have to do anything to set it up. They're going to take the calls. And then they're obliged to take your call and listen to your pitch afterwards.

One more sales tip: Stand up! Put a smile on your face, get excited, and move around a bit while you're calling. Motion creates emotion, and emotion creates connection and rapport. These are small-business owners, their time is valuable. So make it quick and get to your value proposition.

You can build a relationship, sure, but just get to the value proposition. If they're not available, ask their name, put it in your spreadsheet so that you know who to call back and ask for next time.

You can also approach them through email *after* you've called them. Send them information and case studies. Just show them how you can help them. And always remind them you want to send them five or 10 free calls.

Make sure you review the results based on the call quality to be absolutely sure they're getting business. Also check to see how many calls they answer. If they let a lot go to voicemail, they're probably not going to be a good buyer. You'll know to watch them or maybe raise the price.

You may not get them the first time you reach out, but you will get them if you keep trying. If you can help someone's business grow, eventually they'll listen to you and eventually they'll become your client.

Working with buyers

Now, I'm going to say this again because I really believe that it's the way to do this: Give away calls for free to establish a track record. When you go to the store on free sample day, do you take the free samples? Yeah. Some people go to Costco on free sample day only for the free samples, and then they end up buying a bunch of stuff. Because, as humans, there's the power of obligation. Someone gives you something for free, and you feel obligated to reciprocate unless you're a sociopath. And you're not. I hope.

For 95% of the population, if you give them something for free, they feel obligated to reciprocate. If you give them five free phone calls that make them thousands of dollars, they're definitely going to take your call back. They feel like they owe you. Giving it away for free is more than just your foot in the door, it's almost a guarantee of success.

Once you sign them on as a buyer, bill them in advance. Don't issue credit for two reasons. One, people don't pay their bills. Just because a guy has a beautiful dentist office with a grand piano that plays itself and a waterfall and a 300-inch TV and a Bentley parked out front does not mean that he pays his bills. Maybe his office manager sucks. Maybe he's got a ton of debt. I don't know and it doesn't matter. Some people don't pay their bills. What you're selling is worth its weight in gold, so make them pay you in advance.

The second reason is if you issue people credit, then you've got to go collect. You've got to invoice every single buyer, and follow up, and blah, blah, blah. Giving people credit is not a good idea unless you hook up with a dentist office that has like 12 locations. For that guy? Sure. After they prove themselves, you can have your accountant work something out. But if you're dealing with a guy who buys four calls a week, do not issue credit. It will be nothing but a headache.

Now, you want these people to work with you for years, right? That's the whole reason you're doing all this work up front. So help them be successful. It's not enough to just drive leads to their door, you have to help them turn those leads into customers. If you think that's not your job, you're leaving a ton of money on the table.

Listen to a new buyer's calls for the first couple of days. Record their conversion rates. Take notes on their pitch and coach them to do better. Give them pointers. Become a coach for a while. And yes, do it for free. Because the higher the close rate, the more they'll bid for those calls, the more you can create upward competitive pressure. All that leads to more money in your pocket *and* a client you love working with because they never hassle you. And if they spend $2,000 a week with you for five years, that's a lot of money.

All you have to do to do right by them is a little bit of quality assurance and schedule follow up calls to make sure they're selling effectively. Your effort in training your customer is going to pay itself back in dump trucks of cash.

As you go through this process and learn what works for you and your clients, create a best practices guide. This is just a couple of pages in a PDF that gives them examples of what to do, the best way to close customers, and the best way to be more successful. Then you can use the guides as a teaching aid so the whole onboarding process becomes a repeatable system.

OK. You've taken lots of notes. You've documented your scripts. You have every single call recorded. You've created a CRM to keep track of your accounts. Now take those recordings and scripts and build a training program for people to find more buyers for you. As I mentioned before, you could build a whole call center full of people who do nothing but grow your buyer network.

Then once you've conquered Independence, head to neighboring suburban areas. Then go statewide, then go regional, then go national. Eventually, you will have a national network of call buyers in a specific industry, and you can sell your business for so much money you won't even know how to spend it. Which, in my case, means I get to start all over again in another industry because it's just so much fun!

Leverage affiliates to scale
Once you understand all of the metrics, you have all of the buyers, you know all of the numbers, you can bring on your own affiliates to scale your own business. Now, this may work for you or it may not. This is an entirely different way of going about things, but that doesn't mean you can't do it. It just requires learning. You create a list of the offers you've built out and post them in groups, on forums, or reach out to your affiliate friends. Tell people you have an offer that's converting, and you need more phone calls.

If you're going to do this in a big way, you'll want to hire some affiliate managers, a quality assurance team so that you can stay on top of your affiliates, and an accounting and billing solution so that you can pay them. Start slowly with affiliates so you prevent issues. You can build a whole team around supporting your business. And if you've used Ringba to create your own network of buyers through load balancing calls, you now have your own offer. By the way, this is why networks will often tell you that you don't need your own tracking, because you can build your own offers and won't need them anymore.

Start your own network
Once you have some affiliates of your own and things are running smoothly, you may decide to take it all to the next level and create your own network with lots of buyers and affiliates in different verticals. This model is not for everyone. I personally wouldn't do it because you're spinning plates

all the time. And if you don't have a distinct competitive advantage in a vertical, it's hard to run a network. But some folks love it, so let's walk through the process.

First of all, I highly recommend calling your enterprise an agency instead of a network. An affiliate network has a negative connotation with the big brands. The word network just says *I broker deals*, and big brands don't really like that. It's not the network's fault, it's just the ecosystem. Treat it more like an agency; companies that invest in their brand will feel more comfortable working with you.

Then, since you started as an affiliate and you have your own media buying division, you're generating your own traffic. Brands prefer that to a network that relies entirely on its affiliates. They want to see that you have your own traffic because it shows you understand how the pay per call game works, that you understand consumer intent, how the backend of a campaign works, and how important quality calls are to them.

Your network should exhibit at trade shows—lots of them. Most networks don't do this, but the few that I see at every show are growing rapidly. It's a lot of work and it's costly, but you can build an entire business from just trade shows. You're building a reputation as a pillar of the community. We spend a ton of money at trade shows every year for Ringba, but it's really important to our growth.

You should also sponsor events. It shows goodwill, and people learn who you are. When people know who you are, they're willing to take your phone call. Name recognition opens doors.

The big game here is to leverage your own hard work into growth. If you're going to build a network, focus a ton of time on quality assurance. I don't know a single network that puts in enough work on QA anywhere, and I think that's a detriment to their business. It's not overhead, it's a profit center when done correctly.

Lastly, and this is the most important part, be careful with payment terms. If you pay too quickly, you will get defrauded and be left holding the bag. Never ever give new affiliates uncapped offers. Hourly, daily, weekly, monthly, and concurrency cap every single affiliate from now until the end of time! New affiliates need time to prove themselves, and a surge of bad or fraudulent traffic can destroy your buyer relationship. Give yourself time and a runway to QA and assess new affiliate traffic without risking all of your hard work. Even after an affiliate proves themselves, always cap them even if those caps are 5,000 calls per day, because it protects you from spam attacks or an affiliate going off the rails.

Always communicate the caps to affiliates whenever you change them and make sure to let them know you pace all new affiliate relationships to determine quality. If you're clear about this, it will scare away the problem people. Pay new affiliates slowly, make sure you get paid before you pay other people, and make sure your payment terms are in line with the amount of capital you have to lose.

Vetting new affiliates

Just like the networks checked you out when you were new, you need to do some due diligence and vet people who want to be your affiliates. Check their articles of incorporation and make sure they're in good standing. Always interview them over a video call to verify who they say they are. People will create fake Skype and social profiles, and try to get in without ever speaking or doing a video call to trick you into working with them. (This goes for buyers, too, by the way. Always do video calls.)

Google their company names with the words scam, fraud, and rip-off. Check Better Business Bureau ratings; check addresses to see if they're real using Google Street View; check IP addresses to make sure they're in the right country; ask for references and actually call them; check social profiles and reverse image search profile pics to make sure they're real. Do your due diligence! Just 15 minutes can save you $15,000 in losses and major reputation damage. We've got a checklist you can follow inside the Pay Per Callers forum. Download it, use it, and save yourself some major headaches.

CHAPTER FIFTEEN

NEGOTIATING FOR SCALE

Back in the day—yes, in my parents' basement—I was negotiating with Yahoo trying to get a discount on ad space. I called the account rep and said, "Hey, this CPM rate just isn't working for me. Like, I can't make any money here." He said something like "Look, dude, you're only committing to $10,000 a month."

I was thinking in my head that I only *had* $10,000, so what was I supposed to do? I told him I was competing against people who were committing to a million dollars a month. How was I supposed to make that work? So I straight up asked him how it worked. What would happen if I didn't spend the million bucks?

His reply was "Oh! No, no, you don't actually have to spend it all. We're all about the commitment, not the actual spend. You just sign the IO (insertion order) for a million dollars and then you can cap it at whatever you want. If you don't get to the million, that's OK. You don't owe us the rest of it. We just do our commissions and forecasting based on the IOs."

I was dumbfounded. "Wait. So you're telling me I can commit to a million dollar IO and cap my first month at $10,000?"

"Yeah, absolutely."

"Well, sign me up baby! One million dollars it is! I'm in!" I faxed over the IO and bam! Pricing adjusted.

Weird, huh? See what you can get by just getting curious and asking questions? I wasn't trying to cheat him, and he wasn't trying to rip me off. I just didn't have a complete understanding until I tried to negotiate.

Negotiation ground rules

Lots of people avoid negotiation because they don't like conflict, or they're afraid they'll walk away with less than they came with. Or they feel like everyone should be equal and that negotiating for themselves is somehow negative. But that's not how I see negotiation at all. If you're providing more value, you should be paid more. Plain and simple. And advocating for yourself is never a bad thing.

Negotiation is important and should happen in just about every aspect of your business. You'll develop your own style, and you don't have to do it exactly like I do. But I want to show you how simple it can be. Before we get started, let's set a few ground rules for negotiation.

1. The most prepared party almost always wins in any negotiation.

In the Yahoo scenario, I was underinformed, but that worked in my favor. Generally, though, if you know more than the person on the other side of the table, then you're in a better position to get what you need out of that negotiation. Unfortunately, in a pay per call context, there will be times when you just can't be the most prepared. You won't always have as much information as the other person, especially if you're dealing with a network or broker. There are some things you can do to even the odds, and we'll talk about those shortly.

Next, know your numbers. You need to know what your calls are worth, how many calls you have, what your potential for scale is, and how much time you'll need to scale your campaigns. Knowing those numbers will give you a baseline for negotiating higher payouts.

If you're a higher-volume partner or you have the potential to be higher volume, it's pretty simple to negotiate preferential treatment with networks and brokers. But you need to present yourself as a professional, otherwise they're not going to take you seriously. You want to be seen as a growth partner, and that only happens when you demonstrate how valuable you can be for them.

2. Try to create win-win situations as much as you possibly can.

Harrison's dad always used to say leave 5% for the other person. Leave some meat on the bone for them. Everyone should walk away from the negotiation happy. Just because someone is a pushover and gives you everything you ask for doesn't necessarily mean you're winning. If you were overly aggressive, and they don't feel good about the concessions they've given you, then they may just decide not to work with you at all. You should be concerned about that. If they feel like you took advantage of them, they may be resentful and shut down your communication channels. That means you lose valuable inside access. You don't want that.

Always focus on creating a win-win for everybody. You can even casually ask them to make sure. "Hey, how are you feeling about this? I think we've got a win-win here."

Once you've sent some calls through to a network or broker, it's a simple matter of politely asking for more money. You just talk to your account manager and say something along the lines of, "Hey, I've been sending you a bunch of calls lately, and it seems like you guys like the quality. My margins on this with you guys are really low because I work with some other partners. I'd love to scale this campaign for you, though. Could you manage a payout bump so that I can be more competitive?"

Most of the time that's it. All you have to do is ask. Notice that I framed that ask as a positive for them. If they give me a bit more payout, I can send them more calls because they will have a higher margin than other partners.

3. Put the details in writing just so there are no misunderstandings.

After you talk with them, just send a simple email or text that says, "Hey, just to confirm from our call today—I really appreciate you working with me on this—we're going to bump the pay out on the [industry] campaign to $12.50 starting on the first of next month, right?"

And then they'll reply, "Hey, great call! Yeah, absolutely. I've updated it in the system." This way you have at least some sort of written receipt that you will be getting that increase, so if there's ever a dispute in the future or if they forget, both parties are covered. Network managers talk to a lot of people every day, and sometimes they forget what they promised. Usually, it's a simple matter of reminding them. But it's much easier if you have that backup in writing.

4. Always be willing to walk away from a negotiation.
Any negotiation you're unwilling to walk away from puts you at a serious disadvantage. This is why you need to have other options at all times, so that you can simply leave the table without reaching an agreement. Nothing changes people's minds faster than pulling volume for a few hours or a day. If you're running a couple hundred calls a day and they won't negotiate with you, all you have to do is reroute those calls to another buyer, and suddenly that network's call center is complaining that their calls dried up.

If that happens, the network will be changing their minds really quickly. But don't be a jerk if you have to walk away. Don't burn any bridges. Just withdraw your traffic and see what happens. If you know your numbers and know exactly how much that loss of traffic will hurt them, you're in the power position. Sometimes, they just have to see it demonstrated.

5. Never make the first offer if you can help it.
That keeps you in a position to agree or disagree. If you make a suggested pay bump for an extra dollar a call, they may have been willing to go up to $3 a call without a fuss—and you just lost out on a huge amount of money. The networks will ask you how much you want. And then try to split the difference between that and what you're being paid now.

So, you always want them to start with their number. You might say, "Hey, I'm driving a lot of traffic these days, and it's converting really well. I'd like a pay bump. How much could you give me?" That positions you to take what they offer and negotiate up.

Sometimes you get stuck, though, and they straight up ask you *how much do you want?* Instead of answering with a number, you can deflect the question by saying, "Well, as you know, the more I can spend on advertising, the higher the volume I can do for you. So, if you have capacity available, I'd really like to understand the best you can do on this payout bump. Then I can go back and recalculate how much of a volume increase I can do based on the higher payout."

Notice this is still positioning your pay increase with the benefit of more volume for them. Still working on a win-win here.

This is why I really like negotiating over live chat or text message, because if you're on the phone with someone who's a fast talker with more information, they may get you to agree faster than you want to. Using chat allows you to do quick research in between messages and lets you think through and proofread your reply. It buys you more time. You're still looking for the win-win, but sometimes you need a few minutes to think. And if it feels like you've been absent from the chat too long, you can always apologize and tell them you had a call or an interruption of some kind.

6. Always have a backup plan in case things don't go your way.

If they don't give you what you want, you need to know what steps you're going to take in advance so that you can execute on them. You never want to walk into a negotiation where there's only a plan A and that's it. Are you going to walk away with your traffic if they say no? Or will you just go back to business as usual?

You should know that if you pull away or reroute your traffic, a lot of account managers lose their commissions. Their pay is directly tied to you in some way. So if your traffic disappears, well, they don't just look bad, they lose money. And since you're following all the advice in this book, you're a serious growth partner that they do *not* want to lose! If you know your numbers and can reroute traffic with the push of a button, you have the power in the negotiation, and they know it.

Network vs. network

The whole reason you're in the power seat with pay per call is you have 100% control over everything you've built. Traditional affiliate networks hold all the cards and if you go away, it really doesn't affect them at all. There are tens of thousands of other affiliates to take your place. Worse, they can pull an offer at any time, and all your work comes crashing down. The only choice you have is to start over.

In pay per call, you keep everything you build. And the better campaigns you build, the more the networks and buyers will compete against each other to keep you happy. It's just a much nicer situation to be in.

Because you're holding the cards, you can leverage your hard work by getting the networks to compete with each other for your traffic. You have to be a little bit careful with this because the pay per call networks all work together in some way, shape, or form, but they are also competitors. So, you can pit network versus network by name dropping, and the likelihood that they're going to go verify that information is really low.

There are a couple of ways to approach this. The simplest is blind price negotiation. If you take your current payout at any network, no matter what it is, and add 10% to it, I can pretty much guarantee that you'll get the pay bump. That's the bare minimum you should ask for. The networks are averaging 35% to more than 100% margin. So a 10% bump is totally reasonable for a valuable partner. If anyone offers you less than that, it's a terrible offer and you should not take it.

If you just ask them nicely, they'll most likely give you at least 10%, and then you can just thank them and keep strengthening your relationship with the account manager. Maybe you get your volume up a little bit, and then you push for more a week or two later.

You're going to want to continually negotiate with networks, especially as you gauge your quality and the volume goes up. This is a volume-based game, so the bigger you get, the more buying power you have and the more money you can demand.

Another tack you can take is to go to network A and tell them that network B also has the offer, and they're paying you more. Of course, tell the truth. If you're bluffing, at least make sure the other network is actually running that offer or something in the same vertical. Many networks do business together and can simply ask to call your bluff. Be prepared and know your numbers. Make sure you really do have multiple networks and relationships all over. When you can demonstrate that you know the value of what you're doing, everyone is going to want to work with you.

Even if you don't have a relationship with network B yet, you can start one by simply saying, "Hey, network B, I'm running 100 calls a day with network A on this campaign. Can you beat their price?" They're probably going to say yes, and then you open up a door and have another network to go to.

Now, if you really want to find out what the wiggle room is on a blind price, you can come back to them and say, "Look, network B offered me 60% higher at the same duration." And then see what happens. Worst case scenario, they're going to tell you no. But, by doing this, you're going to judge the account manager's reaction. If they come back with a counter-offer of 70%, then you have to wonder what the network's margins are actually like. There may be even more room for significant pay increases in the future.

The second type of negotiation you can use is called research price. This is essentially blind price, but you do a lot of research up front to increase the leverage you have in the negotiation. When I used to do negotiations with affiliate networks many years ago, I would always use the research price route because the more you know, the more powerful you are in that negotiation.

My goal was always to know more than my account manager or his entire team about the offer I'm working on. And that's not necessarily hard to do because a lot of these networks are holding 50 or 100 different offers, so the likelihood that every account manager is a master or an expert on every single vertical or offer is pretty low. By knowing everything front to back in the entire space, including where all the other offers are at the other networks, and what they're paying out, and where the rivalries are, you can really position yourself to get the best price on every offer.

And again, if you're being paid more than anyone else on a particular offer, you are in a better position to buy media and market that offer because you can spend more money per acquisition than your competitors and still make the same margin. Your goal isn't to get these price bumps just so you can make more money. It's so that you can increase your volume and then add more assets and value to your business.

With the research price model, you want to speak to three different networks. Tell them, "Hey, I have calls in this insurance vertical, and I am considering running them with you. I've been running this campaign for months with really quality traffic. The other networks are happy with it. What can you offer me?"

That's it. Let them make the first offer. If they say, "Well, what do you need?"

You say, "I'm offering you my business from another network. If you want it, you have to make the offer."

And in that position, you can just be aggressive about it. Don't give them a choice. Make them give you the first offer. They'll do it. They're not going to be like, "No, I don't want a shot at my competitor's volume." That would be absolutely crazy, and they're not going to do that.

And so whatever that network offers you, you can assume that you can get at least an additional 10% if you push them. Whatever they offer you, plus the additional 10%, is your baseline. Then go back to your current network and that's the new minimum they should give you.

Realistically, though, you should tack on an additional 10% above that baseline. Now we're getting into some margins where you have competitive advantage. And that's why you need to speak to at least three networks and get three real offers. Don't take their first one. Negotiate, and try to get those three offers 10 points higher.

What happens if they say no? Reroute your traffic, even if it's just for a few days, so they notice the drop. In my experience, being pretty aggressive about this gets you what you want. I was never scared to be aggressive about this. Even if you make a little bit less money for a day or two, you can set yourself up for months or even a year of higher payouts if you do it correctly.

If you're transparent and tell your account manager, "Look, I can't turn away an additional 15% volume. I'm going to switch half to this other network and give you the next 24 hours to figure it out. Then, if you don't want to match their price—which is all I'm asking for, a match— then I'm going to have to give them all the business."

It's just that simple. Explain that your phone calls are worth a lot. You need to be paid the most for them so that you can compete. If it's profitable for them, and you have a good relationship, they should agree to it. There's no reason for them not to. You want to be careful with bluffing. Make sure what you're demanding is reasonable, and you'll still reach a win-win for both parties. But definitely push right up to the line so you get the most value. Don't leave money on the table.

I will tell you right now that if you bluff the takeaway and you don't actually take the volume away, you're screwed in negotiations with those people forever. You are never getting them to budge for you again. Never bluff the takeaway. You need to have another target configured in Ringba, ready to go, ready to click and just reroute the traffic so that they don't have a choice. Nothing shows how serious you are like pulling your traffic.

You can be aggressive and still maintain a friendly relationship. Just say, "Listen, I truly understand that you're trying to do what's best for your business. And I need you to understand that, for me, it's a numbers game. If I have more money in my pocket, I'm not going to spend it partying or on nonsense. I'm going to actually up my advertising bid so that I can increase volume—which makes us both more money. So by giving me the extra margin here, you're helping your own company grow. I understand if you can't do that, but this other company is willing to invest in our future together. I would hate to move the volume, but I have to capitalize on opportunities."

Always leave with the door still open. "Even if I have to move the volume away, I promise to keep the door open, and we can talk about other campaigns or other opportunities in the future. I still want to work with you, but on this specific opportunity, my margins are super thin. I need every penny I can get to try to scale the campaign, so let me know what you want to do. Either way is fine, but in the next hour, if we can't come to some sort of conclusion, I just want you to know that I am going to reroute the traffic to this other network."

Now, if you say that to an account manager, they might blow you off. Then, after about an hour, you ping them back and say, "Hey, man, sorry. I had to reroute." Suddenly they'll start paying attention. And they will never mess with you again.

Negotiating with call quality

My favorite way to negotiate is by leveraging call quality. In this scenario, you have to set yourself up so you know so much more about the campaign than your account manager that they're not going to know what to do with you. All the research you're going to do also sets you up for finding and negotiating with your own direct buyers, so it's well worth your time.

Your goal is to see how your traffic stacks up against all the other people on the network. If your traffic converts at a higher rate, you deserve to be paid more money for that traffic. It's that simple. And if they won't pay it, and you know your metrics, then it's not complicated to go directly to the buyer and work out a deal that cuts the network out entirely. Some people aren't going to like that I included this tactic, but it's the law of the jungle. And this industry is a jungle.

I'll tell you right now that most affiliates are not doing this—either because they don't have their own call tracking and they can't, or they just don't want to do the work. Either way, you win. You can hire a virtual assistant to do this, but you will learn so much more if you do it yourself. This is about mastering the vertical for the future, not just getting a short-term pay bump.

Listen to 100 unsorted call recordings. Some will be really short, others will be long, but you need to take the 100 as they come in. If you're not doing QA at the same time, you can disregard any call under a minute. The ones that don't convert aren't what you're looking for. If the average handle time of a campaign is 15 minutes, you're going to want to take a look at every call that's 10 minutes or longer. You can assume that any phone call under 10 minutes or under seven minutes is probably not a conversion. You also don't need to count calls that never got answered or didn't make it through the IVR screening if it's an IVR campaign.

You're trying to determine the actual conversion rate. How many of your calls actually converted for the end buyer? Did they make a sale? Set an appointment? Get a registration? Whatever the goal is, how many of your calls convert? Also note how much they got paid for a sale so you can understand how much the call center made on the transaction. You need to see how much your calls are worth to them.

Once we know the conversion rate on 100 phone calls, we're going to determine the revenue per call. This is a really simple calculation. We're just going to take the total revenue we think the call center made on those 100 phone calls, and then we're going to divide it by the number of phone calls.

Now, take a look at the call center's revenue per call and compare your payout to figure out how much the network is making and how much the call center could pay. If the call center's revenue per call is $300 and you're getting paid $6, the network's either taking a massive margin or they don't know how to negotiate with the call center. The goal here is to figure out what the call center's actual margin is.

You can assume that the call center's operating cost is probably in the range of 15% to 33% of their gross revenue. Everything else beyond that you calculate out your best guess for their profit margin. If you know more than the call center does or you know at least as much or have a general idea of how much money the call center's making, then you're going to be able to negotiate with everyone at a level that they're not prepared for.

Information is power. If you know you have higher-quality traffic, use that as leverage. Networks are not going to validate what you tell them for the most part. You want to be careful about bluffing in these circumstances. You should know your real numbers, but always negotiate. Figure out a way to position yourself so that you're a quality long-term partner. They need to keep you, and if they don't take care of you, they're going to lose you.

If you do that, every single network and every single buyer will give you more money for your calls, which allows you to create a competitive advantage in the space and outbid and out-operate all of your competitors.

OK, scaling is great in theory, but how do you handle all those calls and all those buyers? It's a nightmare if you don't have the right technology. So, coming up we're going to look at automated call routing and how you can set up your business to flow the way you want it to.

Ready? Let's go...

CHAPTER SIXTEEN

RINGBA SECRETS: OPTIMIZING CALL FLOWS WITH AUTOMATED CALL ROUTING

You know those ladies, the ones who ran the old-fashioned switchboards? (I don't know if there were men. I wasn't alive then. I've only seen ladies, so just bear with me, OK?)

Back in the day when telecom was new, these ladies connected people's phone calls manually. They were doing call routing by pulling out a pin and plugging it into another hole. They did all the switching, and it was a highly skilled job. They had to know what they were doing and really understand how to route these calls.

Then automated switching came out and cost a lot of people their jobs, but it was the only way to scale the telecom industry because manual switching was just too slow. So at Ringba, we have automated call routing that utilizes a super-computer cluster and AI to look at all the underlying data and make a routing decision in microseconds. It can take a look at the source of the caller, the keyword data, the ad copy, what buyer is available, who's gonna pay the most for a call—any criteria you want. It can even use AI to predict where a call is most likely to convert. It's amazing!

So pay per callers who aren't using Ringba are like those ladies. They're great. Everyone loves those ladies. But they're slow. It's an antiquated system. And you can't maximize the yield of a phone call because those ladies can't see who's available on the other end of the line.

They can't see how much money the person calling in is worth. Do they have four cars? Did they call from a website or from a QR code? Which one? They can't tell any of that because all they have is their little switchboard, and they have to manually adjust the routes and priority.

The top players in pay per call are all using technology to maximize their revenues. And since you want to be one of those top players, I'm going to outline exactly what technology you need to be using and when. You won't start out using all of these, but you do need to know what you're building toward and what your next level looks like. Each one of these features can be used either alone or in combination with each other. Once you get going, you may find that all the features are used on every single call route.

Ringba isn't the cheapest option out there. We charge a little more than everyone else, and we do that because we have the best support team in the industry standing by to guarantee our customers have more success with us than anywhere else. Our technology helps our clients make the biggest ROI possible on their ad dollars and their business investment. In order to do that, we hire more engineers, we spend more time getting the software right, and we invest more money into innovation to push the industry forward. I know I'm biased, but I know this for sure: Work with us, and you'll make the most money possible. Cheaping out on your software is the best way to fail.

Now, if you're not using your own call-tracking platform, you're not going to be able to prioritize and weight your call flow so that you can decide where your calls are going. You'll have no control over the situation. So in these examples, I'm just going to assume you have Ringba so you can see how to control the flows.

Prioritizing and weighting the call flow

The first thing we're going to talk about is prioritizing and weighting your buyers. For whatever reason you may want to prioritize certain buyers and lower the priority of others. Maybe some are payment risks, or maybe some have really high error ratios and drop a lot of calls. Maybe some of them don't convert as well. Or maybe you just have better relationships with some of those buyers, and you really want to send them more calls because you see other opportunities with them.

Here's how it actually looks when it's implemented. You, the publisher, will get your tracking number and put it into an advertisement somewhere, most likely a mobile ad. A customer sees that advertisement, picks up their phone, and dials the number. That call is sent through your call-tracking platform, and the platform prioritizes the route of that call.

MAXIMIZE ROI BY PRIORITIZING & WEIGHTING THE FLOW OF CALLS

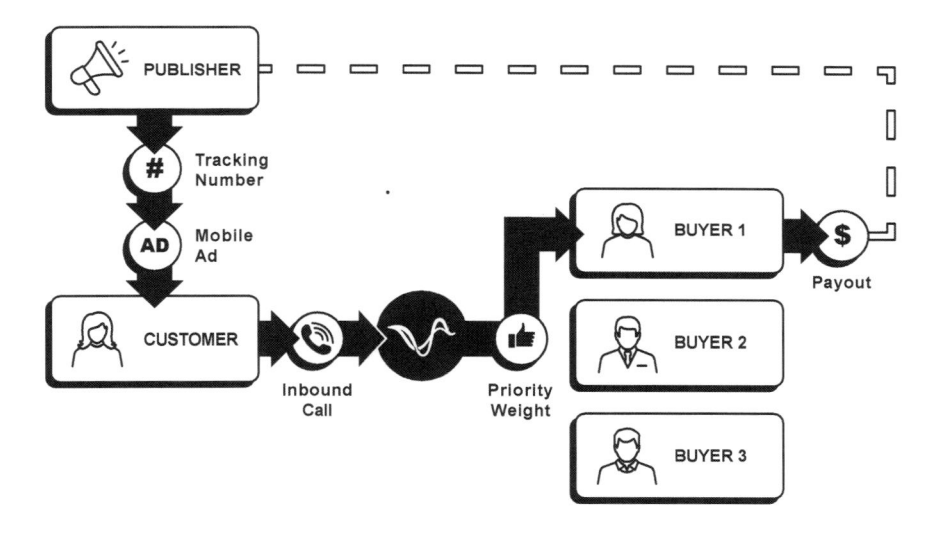

Now, in this simple example, buyer 1 had a higher priority than buyers 2 and 3, and so they got the phone call. The call reaches a duration and the payout comes back from them to you. Whether there's a network or a broker in there doesn't necessarily matter for this example. What matters is that buyer one was prioritized over the others.

Automatically reroute busy buyers

So what happens when the customer gets a busy signal and the call doesn't get answered? Maybe all of the agents are on the phone, or maybe they just are working with a lot of different publishers and are overflowing their capacity on purpose. Whenever an agent's not available, many buyers send a busy signal so they don't have to pay for the call. This is pretty common. When you're working with a call center buyer, they want as many phone calls as they can possibly get at any given time. And they want every single agent they have on the phone all the time because it maximizes the amount of money they make.

One of the games a call center will play is they tell the networks, brokers, and direct publishers that they have more capacity than they actually do. That means their agents are always full. But it also means some calls don't get answered. Now you're paying to get that call, so you want to have an alternative if there's a busy signal. You need some place to send them.

Some call centers allow you to ping them for agent availability. When this option is available, always take it. The Ringba team will help you set up the integration; it ensures that your calls won't sit on hold and die.

When you have call tracking, it's not an issue at all. The system just reroutes the call to the next priority buyer. If you don't have call tracking, there's nothing you can do. You just lose that call unless the network has a backup buyer for it.

Here's what it looks like:

AUTOMATICALLY REROUTE BUSY BUYERS

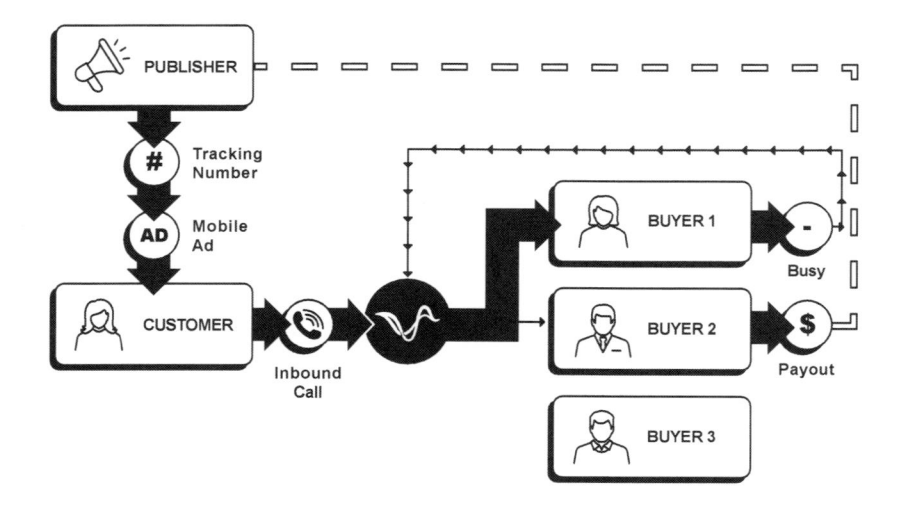

You generate a phone call as usual. The customer calls and Ringba sends it to buyer 1. But buyer 1 sends a busy signal. In a fraction of a second, the platform detects the signal and immediately dials buyer 2.

The consumer on the phone doesn't even notice this. It happens so quickly that no human can detect it. Buyer 2 accepts that phone call. The payout comes back to you. Simple, right?

Reroute unanswered calls

What happens to calls that don't get answered at all? This situation is similar to the busy signal, but the phone just keeps ringing and ringing. If all the agents are busy, they just let it ring in case one of them suddenly gets free and can pick up that call. It might ring 30 seconds or a minute, however long the customer is willing to wait. As long as it's still ringing, the buyer has the opportunity to answer it if an agent becomes available.

If you have a buyer who's doing this, it's really important that you're billing them on duration that starts with the phone call ringing as opposed to duration from answering the phone call. Otherwise, you're paying to send calls that aren't being answered, and that's not good. The buyer is maxing out their agents at the expense of the marketers sending those calls.

REROUTE NON-ANSWERED CALLS

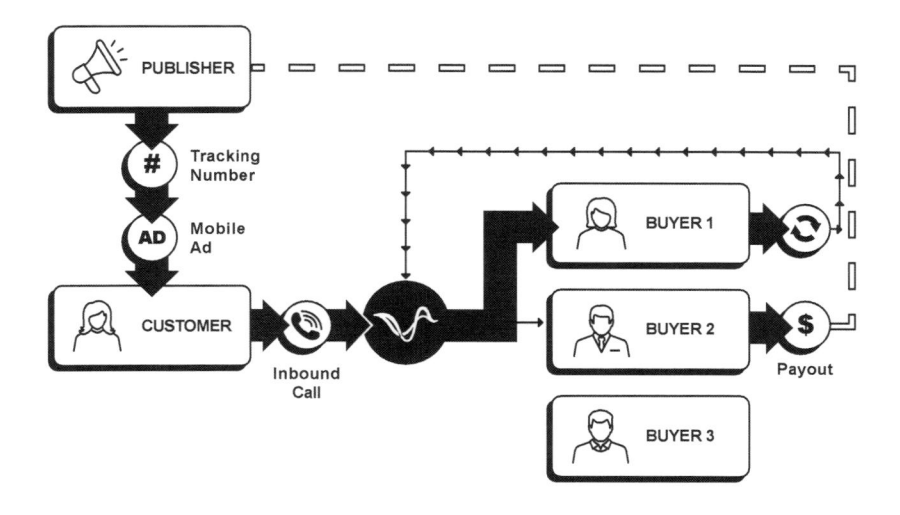

In this situation, you're going to generate your phone call with a tracking number on an ad. The consumer calls, it goes into your call tracking platform, and now there's a timeout associated with the ringing. Inside of the Ringba configuration on a buyer or target level, you can actually decide how long to let the phone call ring before it's rerouted to another buyer for you. Buyer 1 rings for 5 or 10 seconds. Then once the timeout is reached, the call is automatically rerouted to another buyer that may be available. This is one way to recapture the value of a call that may otherwise be lost.

The more calls you have, the more value you will lose by not having solutions like this in place. When you get started, you might be generating just a few calls a day. But you will quickly move to 50, and before long you'll be sending thousands of calls a day. Some of our clients generate tens of thousands of calls every day, and that's really pretty normal in this space. So, if you allow a buyer to just let the phone ring without rerouting, you're going to lose a lot of money.

Simultaneously Dial Buyers and Sell to the First to Answer

Another option is to simultaneously dial several buyers and sell the call to the first one who answers. It doesn't matter how many buyers you have, we're going to dial them at the same time, and the first one to answer the phone call gets it. This is a great way to maximize the number of calls that get answered.

It sounds great, but depending on your buyer requirements and how much each of the buyers are paying, this can actually be a lower return on investment scenario for you. You have to know what you're doing and be a little careful here. Fortunately, on some of the more sophisticated call routes inside Ringba, you can actually simultaneously dial, or "simuldial," a specific group of higher-quality buyers, and then, if they don't answer, you can reroute the call based on how much the other buyers are paying or other criteria. Here's what a simple version of that scenario looks like.

SIMULTANEOUSLY DIAL BUYERS AND SELL TO THE FIRST TO ANSWER

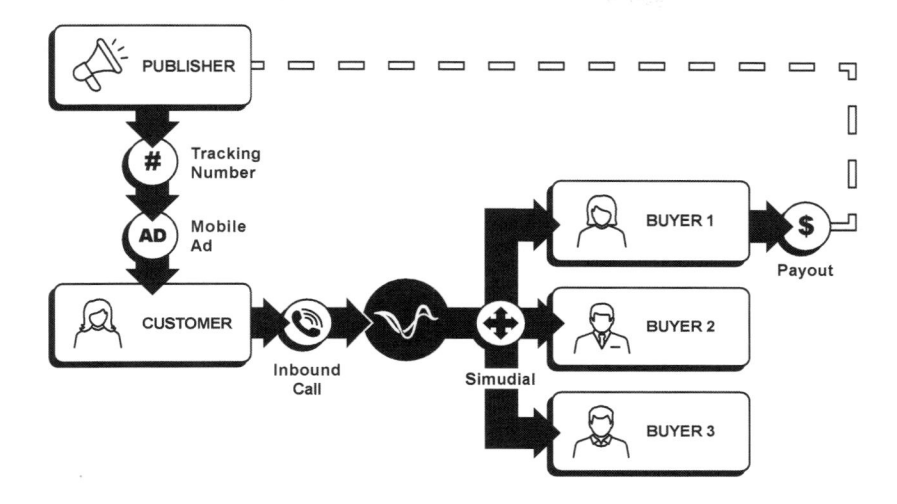

You put a tracking number on an ad. A consumer sees that ad, picks up the phone, and calls. It routes through Ringba. Ringba simultaneously dials buyer 1, buyer 2, and buyer 3. They all start ringing. Buyer 2 may hit it with a busy signal. Buyer 3 may have a technology error. We don't know, maybe they just all ring, but buyer 1 picks it up the fastest because they have a hold queue set up so that they pick up quickly whether there's an agent available or not. Regardless, buyer 1 got the call, so they issue the payout to you once the call meets the requirements.

This can get a little tricky if your buyers have different geographic requirements, or different campaign requirements in general, and you need to route by those. If half of your buyers only take West Coast calls and half take East Coast, then configuring by whoever answers first can be a problem.

Simuldial can result in less money for you as well. If buyer 1 pays $10, buyer 2 pays $22, and buyer 3 pays $6, and buyer 3 is always the fastest to answer the call, your campaign could wind up costing you. Make sure you take into account how much money all of the buyers are actually paying you and what the campaign criteria are before you implement any type of simuldial routing plan on these campaigns.

However, if all of your buyers take calls from any location, you're good to go. Also, if your buyers are all paying you per raw phone call with no duration requirement, you would absolutely want to simuldial, because all that matters is that the call is answered as soon as possible.

Duplicate call routing

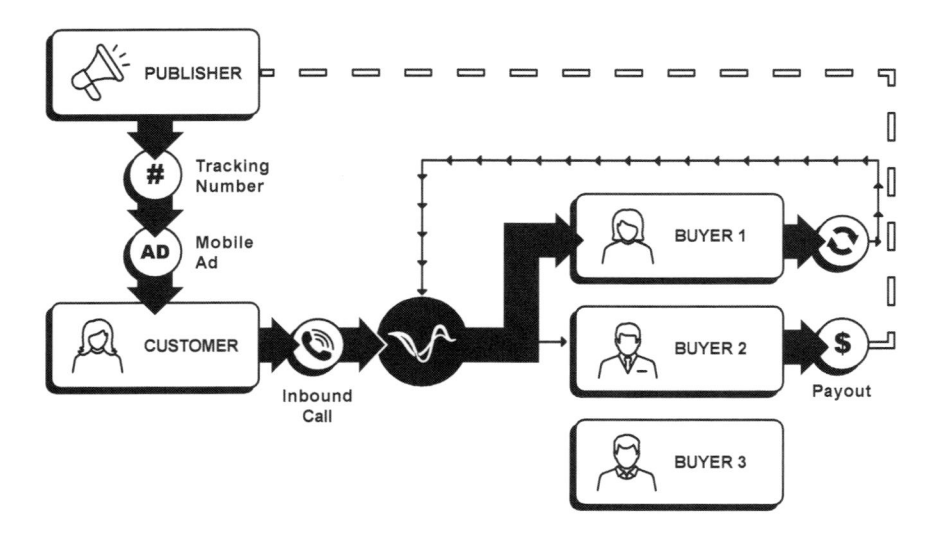

You could make a lot of money by managing your duplicate call flow properly. A duplicate call is when a caller picks up the phone, dials it, calls, they're connected with an agent, then they call again a little while later. If you're working with a network, but you do not have your own call tracking, you would have absolutely no control over this. And networks almost never pay for duplicate calls.

You might think it doesn't happen that often, so who cares? But based on the statistics across hundreds of different verticals and industries, we typically see anywhere between 5% and 25% duplicate calls. Imagine if you could monetize those calls instead of letting the network collect on them. You need to manage your duplicate call flow appropriately, or you're just leaving a ton of money on the table.

DUPLICATE CALL ROUTING

In this scenario, you place the ad with your tracking number. A consumer sees it and dials. It goes through Ringba, and buyer 1 buys the call. Great. But for whatever reason the consumer called back. Maybe they're looking for auto insurance, and maybe they talked to someone at buyer 1, but they were not happy with how they were treated, or the guy they were talking to was new, or the agent didn't do a good job, or whatever. There's an endless list of reasons why someone would call, talk to someone, and then pick up the phone and call the same phone number.

If you generated a legitimate phone call using legitimate advertising, and the call center didn't close that person, that's their fault and their problem. You have a recording of it, because you're using your own call tracking, so if there's a dispute, you have proof that it was legit.

The platform automatically detects if the person's called before and routes them to buyer 2. Now, buyer 2 has no relationship with this caller, they've never seen their caller ID before. As far as they're concerned, it's a fresh call. The call reaches the requirements and a payout is issued. Maybe they sell them, maybe they don't. Again, not your problem, and a payout is issued.

With properly managed duplicate calls, you can actually get paid twice—sometimes more. We've seen duplicate call routing plans remonetize the consumer up to 10 times in rare cases. That means the person who configured their duplicate routing rules is maximizing the value of these repeat callers and then taking the ability away from the broker or network to do this same thing. I think that's the most important part.

If you just use duplicate call routing, and you're working with networks or brokers, that feature alone will pay for your call-tracking platform. Then everything else you do with it, all the other money you make with it, and all the control you get with it, is pure profit. It's gravy.

Geographic routing

What if you need to route calls according to where the caller lives? Call centers aren't always in the same location. Geographic routing simply means we are going to route phone calls by the geographic location of the caller.

You can have a call center that's located in Texas but is licensed to sell mortgages in Massachusetts. It doesn't matter where the call gets answered, only where the caller is from.

Now, most call-tracking platforms will only route by area code. This is a flawed and outdated methodology because people keep their phone numbers when they move. Approximately 35% of people have a different area code from where they're located. That's important to keep in mind, because if someone calls from a 248 area code but doesn't live in Michigan, well, they're going to get connected to the wrong call center. You can do routing by area code, but you're going to burn a lot of your call flow.

With our number pool technology, Ringba is able to figure out the actual zip code, city, state, country, latitude and longitude of where your caller is coming from. You can get that granular. And that means you can maximize your revenue in a whole different way.

GEOGRAPHIC REROUTING

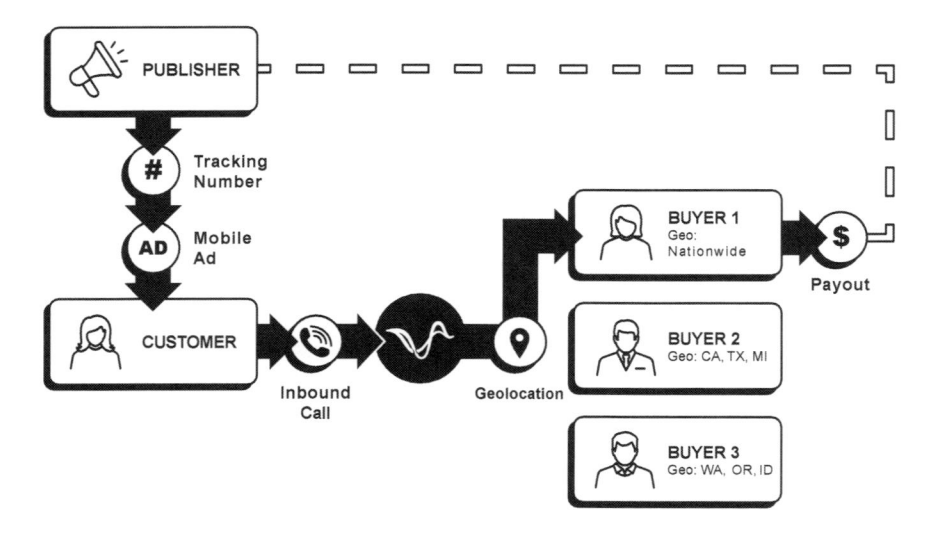

Let's run through this scenario. You create a tracking number, place it on a mobile advertisement, and a consumer sees it. The consumer dials the number, and Ringba makes a routing decision. Now, in this case, let's say it's done by area code. Ringba looks at the area code and it goes, *Oh, it's Adam! The area code says Michigan. All right, I'm going to route it to buyer 2, because buyer 2 accepts Michigan calls.* So far, so good.

But what if I don't live in Michigan? I live in Alaska. The call center will very quickly figure that out and hang up because they aren't licensed in that state. Click. You just burned a call. If you use area code–based geo-routing, you're leaving a lot of money on the table.

A lot of pay per call networks do this because it's easy. They're just brokering. If these calls don't get paid for, that's not really their problem. That's the affiliate's problem and so they don't care. Unfortunately, the affiliates end up eating all this lost value. Worse, they don't have any idea that it's actually happening because they don't have recordings of the calls. They just assume everything is great. But if 35% of the area codes are wrong in a campaign where location is important, there's a 35% chance you won't get paid. That sucks.

So what's the alternative? Ringba is able to track based on the zip code of the person's physical location just by dropping some script onto your landing page. You get the physical real location of the caller so you can drive that to whatever buyer is actually licensed to take that call. Sweet! Of course, someone from Michigan could theoretically be calling about car insurance while they're on vacation in Alaska—but that is not nearly as common.

If you're using this technology, you're going to have an almost 100% hit rate on making sure that your calls go to the right buyers, which means your conversion rate jumps through the roof. You're going to make a lot more money when you use actual geographic routing versus area code routing in these scenarios.

There is another workaround depending on where you're advertising. If you're working with Google AdWords, for instance, maybe you have a campaign in every state. It's possible to pass their location info on in the URL and use 50 different tracking numbers—so the person gets shown the correct number for them. This seems inconvenient, but if you want to get your conversion rate up, it's just a little bit of work to get the job done.

Your goal is to create a competitive advantage that you can use to outperform people who either don't have the technological capability, don't understand how it works, or are just really lazy. A lot of times publishers are a combination of those things, which is amazing because it's not hard to compete. You just work a little bit harder, you work smarter, and you win without really doing a whole lot differently than anybody else. You win just by understanding how this works.

Demographic routing

I was at a trade show recently, and I spoke with a woman who runs recruiting for a major university. They were really concerned about how they fill specific demographic buckets. Universities often enroll a certain percentage of people from different ethnic backgrounds. They struggle with this because they don't necessarily understand how to manage their call flow. How can they track inbound phone calls to the recruiting office and sort them by demographics? It's a really complicated problem.

Other demographic considerations may apply, too, depending on the individual campaign. A debt consolidation company might only want calls from people with more than $10,000 in unsecured debt. A real estate company might only want people with a high net worth.

The demographics can apply to anything. It can be rich people, poor people, or people who work in different types of jobs. It can be different age groups. It can be people who read different magazines or have different hobbies. We're talking about a really wide assortment of qualification criteria.

DEMOGRAPHIC REROUTING

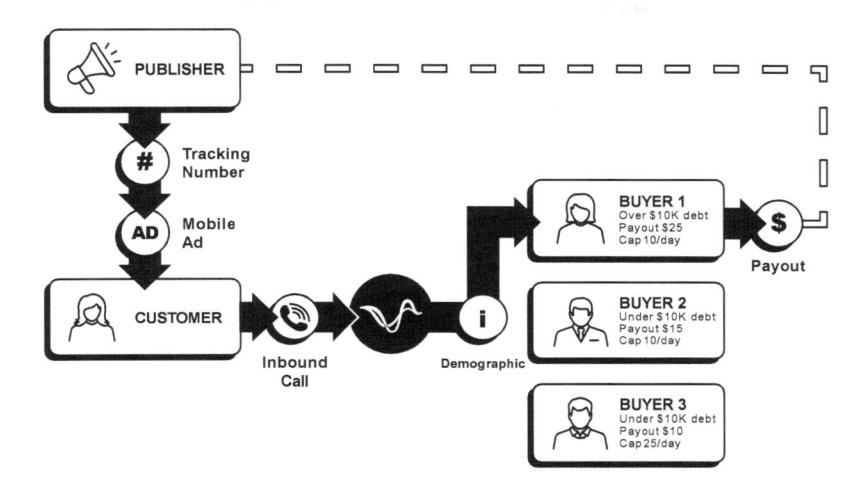

Let's look at the debt consolidation example for a moment. You send out a mobile ad with your tracking number on it, the consumer dials the number and is presented with an IVR. "Thank you for calling Debt Busters. Please press '1' if you have more than $10,000 in debt. Please press '2' if you have less than $10,000 in debt."

If they press 1, cool. They go on to the buyer, and you get the payout if the call meets the requirements. But what if they press 2? What happens if they have less than $10,000 in debt? Lots of pay per callers either won't know what's happening or they'll just let it slide because they have no control of the situation. But if you have the right technology, you can actually reroute that call to someone who will take a different demographic criteria.

There's another element that can come into play here as well. Buyers will do something called capping. That means they are capping the number of calls they will buy in a certain amount of time. Maybe they'll buy 10 calls a day or 20 per month. Whatever the cap is, once they've reached it, you want to have the technology automatically reroute to yet another buyer who's available to take the call.

You never want a situation in which you have callers and nowhere to send them. So inside Ringba, you can set up a waterfall of options based on anything you want—region, IVR response, demographics, payout rates, conversion rates, whatever you like.

If you're using our number pool technology, you can get really sophisticated with demographics. Say you're creating social ads on Facebook. We all know Facebook has a ton of demographic data on all its users. With number pools, you can actually pass that data to your landing page and make routing decisions based on that data.

If you have an advertisement that targets low-income people, for instance, you can actually flag that, tag it as it comes in to Ringba, and then route by that information. If you're running specific advertisements like the university example, where you need to know the ethnicity or the background of the person, you can then pass the lead data into Ringba. On the landing page they'll see *Hey, you qualify for XYZ university. Please give us a call.*

The person picks up the phone and calls. We know what ethnicity they are and can route them based on priority. If that university specifically has a need for a certain ethnicity, they can prioritize that phone call, make sure it gets answered.

When you're working with less sophisticated networks, they'll offer you fewer options. Usually because they use technology that does not allow waterfall payments. It's either $10 for a call or nothing. They don't have a lot of options. But Ringba networks can actually pay you based on call criteria. They can have 25 different payouts for the same call flow based on demographic information, geographic location, length of the call, all sorts of stuff. Understanding how to route by demographic information, and figuring out which buyers are willing to pay more for certain demographics can literally make you a fortune if you do this properly.

IVR selection/qualification routing

Remember, IVR is simply interactive voice response. It's one of those prompts that you get when you call a company. The recording says something like *"For sales, press '1'. For customer service, press '2.'"* Because they want to filter off the customer service calls and route the sales calls directly into the sales call center.

IVRs can be used for a lot of really cool things. Ringba has some incredible IVR features built into our drag-and-drop interface. You can create an extremely complicated IVR in our system in a matter of a few seconds. We also have text to speech in 15 different languages, male and female. If you want to take your campaigns overseas, or you really want to make these things fast without going out and doing a bunch of professional recordings, our system allows you to do that.

IVR SELECTION/QUALIFICATION ROUTING

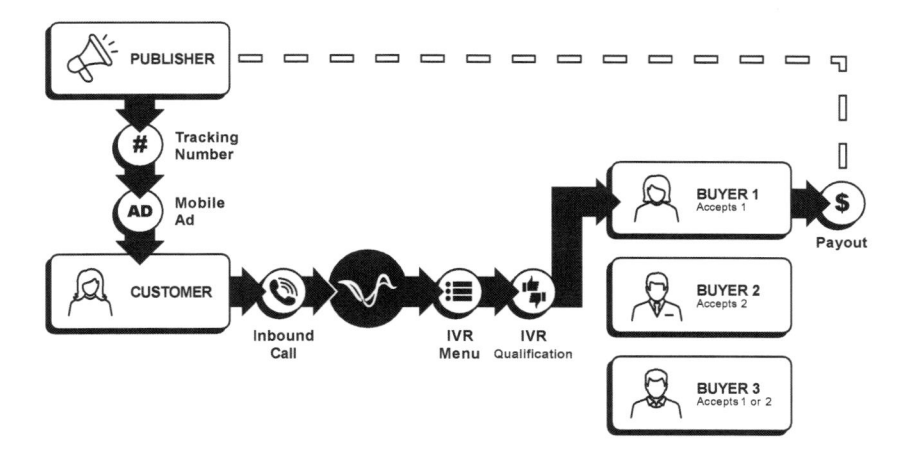

[Diagram: Publisher → Tracking Number / Mobile Ad → Customer → Inbound Call → IVR Menu → IVR Qualification → Buyer 1 (Accepts 1) / Buyer 2 (Accepts 2) / Buyer 3 (Accepts 1 or 2) → Payout]

Let's say you drive a phone call, it routes through Ringba, and hits an IVR menu. *Thanks for calling. Please press "1" for sales, press "2" for customer support.* They press 1, it goes to one of these buyers. Bam! Buyer 1 gets it, and makes a payout. That's a nice simple IVR example.

Now in most cases, there's going to be more of a qualification process. If we use the debt consolidation example that we did earlier, that IVR is going to sound like this: *If you have more than $10,000 in credit card debt, please press "1." If you have less than $10,000 in credit card debt, please press "2."*

In this scenario, buyer 2 wants people with less than $10,000 in debt. The other ones want people with more, or buyer 3 will take either and will most likely pay you different amounts of money based on what that IVR qualification is. With Ringba, you can sort first by IVR answers and then put people into a separate pool that prioritizes based on payouts and capping. There's no limit to the ways you can set up your call flows.

Sometimes with IVR scenarios, you qualify the customer, but then the call center also has an IVR. The poor caller has to go through the selection process twice, and that's just going to piss them off and drastically lower your conversion rate. So, in the tracking platform you can automatically dial the first IVR selections into the second IVR so the users don't even hear the second set of questions. Ringba automatically connects the call to buyer 1, automatically makes the IVR selections, and then the consumer is connected. You get the higher conversion rate because the consumer did not go through multiple IVR trees.

I cannot tell you how many times I've listened to a call recording during which the user is forced to go through multiple IVR trees that are exactly the same. Don't force your consumers to do that. Pay attention to what's going on. Listen to your call recordings. Understand what's happening on your call flow.

Routing based on hours of operation

You can also route calls based on when the buyer is open.

ROUTING BASED ON HOURS OF OPERATION

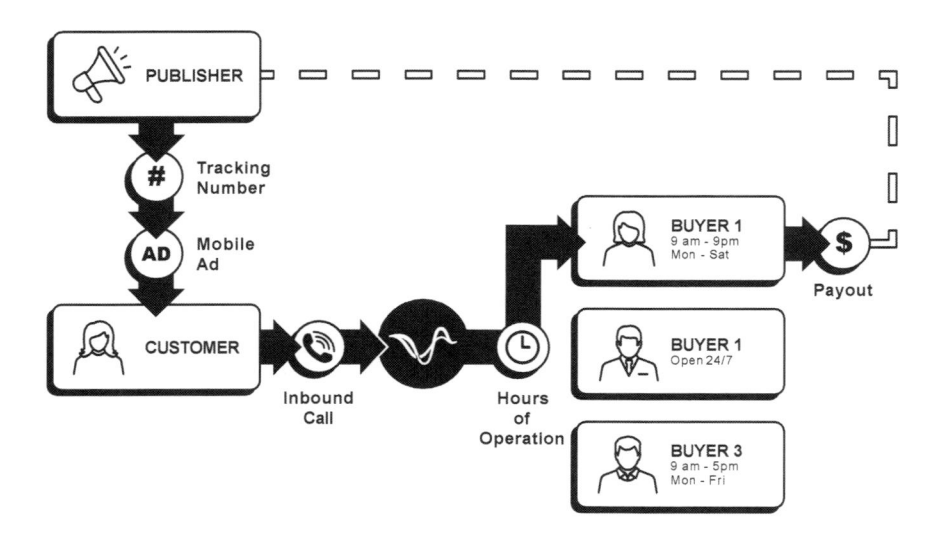

You generate your tracking number, you place it on an ad. A consumer picks up the phone and calls, and Ringba takes a look at the hours of operation of all your buyers. If any of those buyers are closed, we don't send a phone call to them. Because if they're closed, theoretically, they're not going to answer it.

I say theoretically because one of the ways a network can make money is to tell you the hours of operation are for 8 a.m. to 6 p.m. when they're really 8 a.m. to 6:30 p.m. Then, if any calls come in after that 6 p.m. cutoff, the network still gets paid but you do not. It's called arbitraging the hours of operation. And it's yet *another* reason why you want your own call tracking—so you get paid for all your calls, not just a few.

Let's say it's 1 in the morning and someone calls in. Buyer 1 is only open from 9 a.m. to 9 p.m. Monday through Saturday. Buyer 3 is only open from 9 a.m. to 5 p.m. Monday through Friday. But buyer 2 is open 24/7, so it doesn't matter what the payout is. If everyone else is closed, the call goes to whoever is available and open. A payout is issued, and you get paid. Simple.

Routing based on concurrency

When someone talks about call flow concurrency (cc), they're talking about the number of simultaneous phone calls that a buyer can handle at any given time. If a publisher says, "Hey, I've got 50 cc of auto insurance phone calls." That means they have 50 people at all times on the phone looking for auto insurance. If a buyer says they can handle 20 cc, that means they have 20 people available at any given time to take 20 simultaneous phone calls.

The more brokers or networks a call flow goes through, the less they are actually on top of their capacity and concurrency. It's a lot to manage. If you're talking to a buyer through a broker that's going through a network, and you're talking to the network account manager, you're not getting real-time information from their buyer. You'll run into capacity issues.

Any time you're working with a network, be sure to ask what the concurrency is. If they tell you they can take unlimited calls, they're full of shit. That's not how a call center works. Call centers have people in them, and they need people to answer the phone. To take unlimited phone calls you have to have unlimited people, which isn't possible.

But the buyers want to be busy 100% of the time, and the networks want to provide that many calls. It's in both their best interests to get lots of affiliates driving a ton of calls. They don't really care that the affiliates driving the unanswered calls don't get paid. That's your problem, not theirs.

If a network says they have 10 concurrencies available on a certain campaign, that's reasonable. They can work in an additional 10 cc because they're probably working with a whole bunch of buyers. If you add all their buyers together and look at the available concurrency on a nationwide campaign, an additional 10 cc is not that big of a deal. If they say 500, you should probably be concerned and ask them about their buyer network. If they say they just have one giant buyer, that's really fishy. Like, they just had 500 call center agents sitting around waiting for your calls? You need to know what you're getting into before you start paying for ads.

ROUTING BASED ON CONCURRENCY

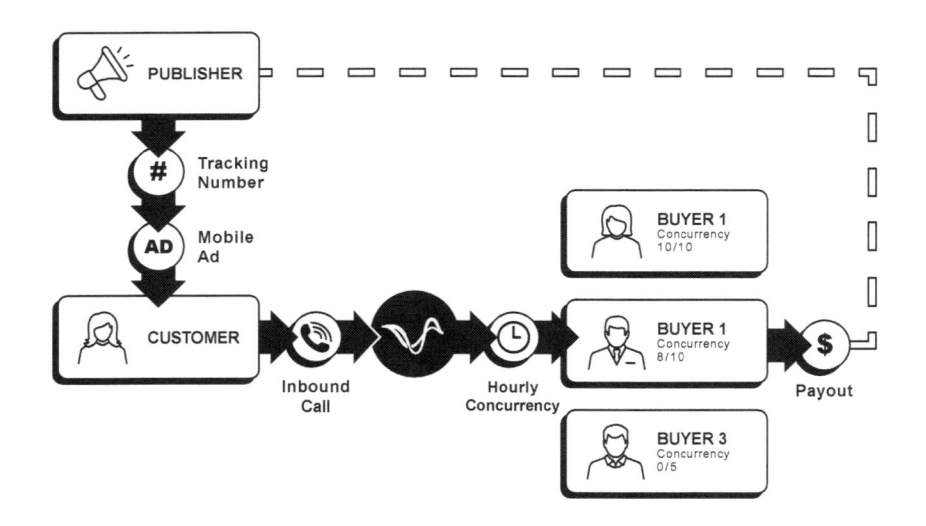

```
PUBLISHER ─ ─ ─ ─ ─ ─ ─ ─ ─ ─ ─ ─ ┐
    │                              │
    ▼                              │
Tracking                           │
Number                             │
    │                              │
    ▼                              │
Mobile                  BUYER 1    │
Ad                      Concurrency│
    │                   10/10      │
    ▼                              │
CUSTOMER ──►──►──►──► BUYER 1 ──► $
                      Concurrency Payout
 Inbound    Hourly    8/10
  Call    Concurrency
                       BUYER 3
                       Concurrency
                       0/5
```

So here's how your own concurrency flow works. You set up an ad with your tracking number. A customer calls it, and Ringba's going to take a look at how many concurrent phone calls all the buyers have. You can configure the concurrency at a flat rate, say they can take 10 cc at all times. Or you can actually configure their hourly concurrency by day, which is really the best way to do it.

Buyer 1 is 10 out of 10—they're maxed out. They're not getting any more phone calls until someone hangs up and is available. Then we'll automatically route them to another phone call based on their priority.

Buyer 2 is next in line and eight out of 10, that means they can take more. They get the phone call, the payout event happens, you get paid.

Buyer 3 gets what's left over. Maybe one and two are your direct buyers, and the network just handles your overflow. Networks usually take a pretty big cut, so prioritize them below direct buyers. But it's still worth having a network or two because you're better off when the call gets answered, as opposed to just letting it die.

The best way to configure your concurrency if you're running calls to multiple buyers is on an hourly basis, especially if you're direct with those buyers. Because call centers operate like any other business, humans have to take the calls, right? What do humans do? They use the bathroom and they eat stuff. At noon, lunch hour happens, and a lot of call centers do not run 15-minute shifts where people are cycling on and off break. They just run the place like a regular business.

If you want to be smart about this, ask the buyer what their schedule looks like? How does your concurrency change? The call center might say, "Oh, noon is lunch hour. We lose 20% of our capacity. Shift change is at 2 p.m., so we get a lot more capacity then. Then at 7 p.m., half the agents go home. We still want phone calls until 11 p.m., but we have half the capacity." Now you have information you can use.

Typically a call center staffs 1.4 people for every person they actually need. People call in sick, they have emergency issues, they've got to leave early, and schedules change. People don't show up. People quit. You want to communicate with buyers as you grow, and as they grow, to understand on a daily basis what their concurrency is so you can properly manage your call flow.

Real-time bidding for phone calls

The world's largest buyers of phone calls buy them programmatically, and that means they're using computers to make decisions on which calls they're going to buy, how much they're going to pay for them, and where their calls are going to go.

Here's how real-time bidding works conceptually. A consumer calls in and Ringba pings out to all of your programmatic buyers. We send information about that phone call, demographic information, metadata about the call, maybe the geographic location of the caller, category of the phone call, whatever information the buyers want. Before the call even starts ringing, we're sending all that information to your buyers.

Your buyers have systems set up to review all of that information automatically. There are no humans involved here. And in less than a second they return a bid, the duration, or other requirements to pay that bid, and any other information that's required for our platform to properly route the phone call.

Then Ringba collects all of the bids, determines who we should sell the call to based on the highest return for you, and the call is routed to that destination. This literally happens before the caller even knows that they're being connected. Most of this happens, including all the pings, posts, and routing decisions, in a couple of hundred milliseconds. (You just gotta love technology!)

And here's what a call flow looks like when you're dealing with real time bidding for calls. A phone call comes in, it hits Ringba's platform, and Ringba pings out to all the buyers. Now, it could be one buyer, it could be three buyers, it could be 3,000 buyers, it doesn't matter. In this example here we have three buyers.

> Buyer 1 returns a bid of $25 and a minimum duration of 1 minute.
>
> Buyer 2 returned a bid of $20 and a duration of 1 minute and 30 seconds.
>
> Buyer 3 returned a bid of $18 and a duration of 2 minutes.

HOW CALL ROUTING WORKS WITH REAL TIME BIDDING

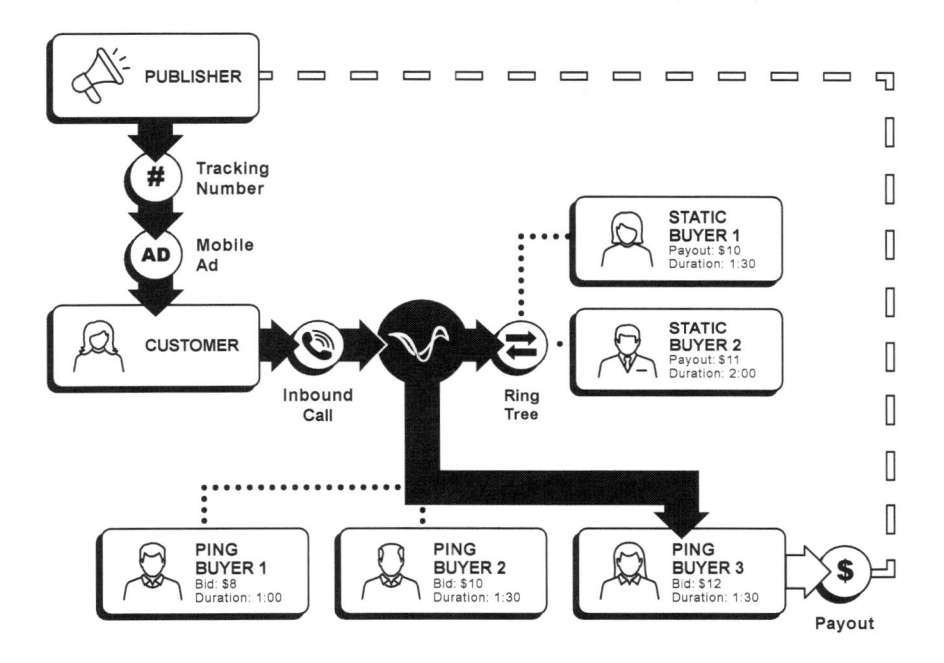

Since the first buyer had the best bid, and you're going to make the most money if the call goes to them, they win. And Ringba routes that call to buyer 1.

All of the other types of routing plans come into play here as well. So, if buyer 1 doesn't answer the phone call, the call dropped, or there's a busy signal, we're going to automatically go through this process again and route it to a different buyer whether they're a ping buyer or in a normal routing plan. We're not going to let that phone call die.

Everything that you're used to with our standard routing also applies when you add a Ring Tree into the fold so that you can maximize the amount of money that you're getting on every single phone call. Here's what it would look like if you have both real-time bidding and static (flat rate) buyers.

You generate a phone call through one of your tracking numbers. Obviously, the consumer sees your ad, they pick up the phone, they dial, it hits Ringba. Then Ringba sees that you have programmatic buyers and static buyers configured for this campaign. So, before the decision is made, we ping out to all your buyers, and they return bids for that phone call.

> Buyer 1 returns a bid of $8 with a 1-minute duration.
>
> Buyer 2 returns a bid of $10 with a 1-minute-and-30-second duration.
>
> Buyer 3 pings back with a bid of $12 dollars for a 1-minute-30-second duration.

Once all of the bids are in, we'll take a look at your static buyers and predict which buyer is actually going to have the highest likelihood of converting for you. Which ones are going to make you the most amount of money? We actually score all of your buyers to see which one has the highest probability of yielding a conversion, and we take that into account when Ringba makes its decision on where your call should go.

In this scenario, buyer 3 is the obvious winner because they bid significantly more than everybody else and have a reasonable duration.

This is a very sophisticated and complicated process, but once you're able to integrate with all these call buyers programmatically, you'll see a drastic lift in the amount of money that you can make per call. You'll also be able to sell calls in thousands of different categories and verticals that pay per call networks just don't have the staff or the technology to manage. You're managing call flows with no humans involved in the process, which means everyone makes more money.

Ringba account managers are able to fast-track you into all these giant programmatic buyers. We can open up massive opportunities for you that none of our competitors can simply because we have better technology, and really good relationships with all the programmatic buyers.

Now, if you're in the pay per call space right now, and you're *not* working with programmatic buyers, I highly recommend you get started now because this is the future. I would honestly say that if you're not using

Ring Trees at this point, you're not even in the game. And if you get into the programmatic space for calls right now, you'll be able to move faster and go farther while everyone else catches up.

Predictive routing

Would you love to wave a magic wand and just know which caller has the highest likelihood of converting with which buyer? Yeah, so would all our customers. While we don't have a magic wand, we do have the magic of technology, and so we built a system called predictive routing using programmatic yield optimization.

Ringba can look at the caller metadata, the target metadata, and real-time conditions to predict which connection is likely to yield the highest estimated revenue per call (eRPC) and pay out the most money to you.

With this system in place, we can adjust if a buyer starts performing poorly. Maybe a caller from Ohio converts best with a certain buyer after their 3 p.m. shift change on Tuesdays. We can get that granular.

You can create your own unique customizable algorithms based on your business goals and attach them to any campaign. You can make as many algorithms as you want and swap them out whenever you want. As of this writing, predictive routing is still in a closed beta testing phase, but we've already seen 12% to 95% increases in revenue per call across millions of dollars of calls involved in the beta.

As you get into the thousands of phone calls a day, managing and optimizing your call flow becomes a full-time job. I don't know a single company running thousands of phone calls a day, or doing seven figures a month in call traffic, that doesn't have someone sitting on a multiscreen setup with Ringba statistics all over it. They sit there all day long, rerouting, managing, and communicating with all their buyers to make sure that they maximize their call flow yield.

That's the fun thing about calls, you actually have to pay attention to them, and if you do it right, that alone is a competitive advantage. You can just start to wipe the floor with people because you're willing to work hard, talk more, and communicate better.

The goal here is to get as many phone calls answered as possible without the caller hanging up so that you get paid for as many phone calls as you possibly can. If it all seems too complicated, don't worry. We've designed our Call Flows visual builder to help you set everything up in seconds. You can even split test IVR options like male voice versus female voice, accents, menu, and scripting.

We see people coming on our platform, learning it, and the smart ones make millions of dollars. I'm so proud of these people. It's an amazing thing to me. It's very rewarding, and it's one of the reasons why I'm writing this book. I want to help more people get into this space and build a real sustainable businesses so they can provide a better life for their families. The more you know about this stuff, the more successful you're going to be.

Some people think they're saving money by not using the most advanced tools available. But can you out-optimize a supercomputer cluster? I know I can't. What happens to these people at the end of the day is they drop calls, they don't route and prioritize correctly, they don't monetize their duplicate calls, and they make far less money. I built Ringba because I wanted to have the most powerful advantage I could get over my competitors. And I want that for you, too.

CHAPTER SEVENTEEN

WHY AFFILIATES FAIL

Back in the day, when I was sitting in my parents' basement trying to figure out this brand new thing called Facebook advertising, I wasn't screwing around. I was in a dark hole in the ground surrounded by painter's tarp, a crappy little space heater at my feet, and freezing my ass off. But I knew—I KNEW—if I stuck with it long enough and tested more times than anyone else would possibly test, that I'd make it.

My whole childhood was spent working harder than anyone else at the things that mattered to me and ignoring everything that didn't (like, um, school …). I freakin' love marketing. I love making ads. I love learning new platforms. And I love competing with myself to see how much better I can get month after month.

That is the attitude I want you to have. Yes, pay per call is a gold mine for those who understand how to work it. But only if you don't give up.

The reality is you need to test, test, test, test, test, test—(deep breath)—test, test, and then keep on testing. Because you will do nothing but lose money in this business—until the very moment that you don't. And then, it's like money just rains from the sky. But so many affiliates and digital marketers just never make it to that point.

I was in a mastermind meeting the other day, and a new guy who had only been in the business for five weeks was complaining about all the money he'd put out and lamenting on why he wasn't profitable yet. I literally stopped him mid-whine and said, "Dude, if you put as much effort into testing advertising as you do complaining, you'd probably see some success."

He wasn't happy with me, obviously. But then, a few days later, he reached out and asked me for help. It's never too late or too early to change your mindset.

So many people fail at this business because they have a preconceived notion about how long it should take them to be successful. They try five or six ads, and they think they've done a lot of work. But the reality is that back in that cold, dark basement, I was literally writing scripts and trying to figure out how I could upload *thousands* of creatives into Facebook at once. Not 5 or 10 ... *thousands*. Maybe I only had 15 or 20 wins out of a thousand. But then I could optimize those few and crush the campaign.

How many iterations are you willing to do? It doesn't have to be slow. You can go at your own comfort level and budget. But you can't have unrealistic expectations. There is work involved here. And part of that work is testing until your eyeballs bleed. But when you win? We're talking about life-changing success.

I've been in the industry a long time, my entire professional career. It's over like 20 years at this point, and so I've seen a lot of affiliates fail and I've failed. A lot. We all make mistakes, but if you're going to survive and thrive here, at some point you've got to grow up.

Most affiliates are in and out within six months, and that includes people who make a lot of money. Affiliates will find one campaign that works, they'll make a little bit of money, they'll spend all of it, and then they don't show up again.

Here's a tip: The affiliate who's all dressed in Louis Vuitton and Gucci and covered in diamonds and whatever and talking about all the money they have probably isn't doing nearly as well as they seem. Whereas you'll see billionaires walking around in T-shirts and jeans because they don't care. It may seem strange, but none of the material possessions actually matter—so don't worry about those things.

Instead, develop financial discipline so that you can grow your business. It's much easier to grow a business if you have a giant pile of cash than if you have a bunch of bling and no cash. If you need to hire someone, or you have a campaign turn and it's no longer profitable, or you lose a bunch of money testing—it's really hard to pay Google with Gucci sunglasses. They only take cash. So please, for the love of all that's holy, save as much money as you possibly can so you can invest back into your business and yourself.

If you find some early wins and start making what you consider a lot of money, keep it. Invest it. Save it for your future. Don't spend it on bottle service and bullshit, please, please, please. I've seen hundreds, maybe even a thousand affiliates make this exact mistake and lose their opportunity to build a sustainable business because they blew all their cash.

OK, enough with beating the dead horse. You're in this for the long run, and you want to work as smart as possible now so you don't have to work at all later. There are still traps you can fall into.

The first problem is not choosing a specific vertical to focus your energy on. Many affiliates will run a dozen campaigns trying to figure out what works. The problem is they don't have the cash flow to bankroll that many campaigns, so *none* of them work. That's usually when they bitch to their friends that affiliate marketing is a scam and quit the game.

If I were going to get into pay per call today, I'd pick a couple of verticals that are very similar to each other and have a similar audience. For instance, if I was going to do medical supplement insurance, which is an older group, I would pick other types of insurance that apply to older people, like final expense insurance, so I could reuse my market data over and over again, and upsell them on different things.

Choose two complimentary verticals and focus on them until they're extremely successful. Only then should you branch out to the next *still complementary* vertical. You worked hard to understand your audience; don't be so quick to throw it all away and start from scratch.

The second problem I see is affiliates throwing $5,000 or $10,000 to some guru for training, but they won't invest in attending trade shows and meeting people who will actually help grow their businesses. The people you need to hang with are all at the trade shows. You need to be there. Meet the buyers, the networks. Hell, come over to the Ringba booth and shake my hand. That's why we're there. To build relationships with serious affiliates who want to grow with us.

Events aren't the only way to create personal relationships. Get in your car or hop on a plane and actually go to the place where they are. Maybe you really want to get in with a certain network. So take one of the affiliate managers or even the CEO out to lunch. Ask them questions. Learn about them and their mission. Build something that will last.

They're going to take a risk on you simply because you put in the effort to go shake their hand and say hello. So few people actually do this, but it's the best way to make things happen and open doors for yourself.

Finally, affiliates suck at planning and setting goals. If you ask an affiliate what their goals are, most will say, "Well, I just need to hustle some money. I'm going to try and make some campaigns work."

That's not a plan.

A plan is, "I'm going to test X, Y, and Z on Monday, and X, Y, and Z on Tuesday. I'm going to try and make this one campaign work. I'm going to dedicate my all to it over the next two weeks, and then I'm going to review the results and see if this is the vertical I should be focusing on."

Here's the formula I've used to build the most incredible things in my life:

- Set a specific goal.
- Make a step-by-step plan.
- Give it a realistic deadline.
- Follow the plan.
- Reassess.

If your goal is to be a millionaire, that's a shitty goal. A much better goal is to make $1 million through pay per call by focusing on digital advertising in the finance verticals over the next 12 months. That's specific, with a deadline and actual steps to follow. If you don't reach your million, then you have to take responsibility and hold yourself accountable for the failure. That's why you always have a chance to reassess.

Did you succeed or fail?

Why did you succeed or fail?

What adjustments should you make to do better with the next time-based goal?

If we do this process right, we're either winning or learning. Failure isn't even in the equation. Only when we fail to learn do we fail to succeed.

Write out your plan, put deadlines on it, figure out what you want to accomplish, and then follow it and reassess at certain milestones. I highly recommend that you're setting daily goals, weekly goals, monthly goals, quarterly goals, and annual goals all at the same time so that you're constantly working toward them, reassessing, and holding yourself accountable. That's going to help you grow as a person, and that's really going to help you build a business.

Most affiliates don't do this stuff.

Most affiliates quit.

And that's why you, my friend, will win!

CHAPTER EIGHTEEN

THE FASTEST WAY TO WIN

I built my business on trade shows. And I don't just mean Ringba. I mean every business I've had in performance marketing. I have built them all through trade shows. There is simply no better way to learn from and connect with the right people. And just to be abundantly clear: If you're not attending the trade shows, you're not in the game. Like, you're not even trying. If you're not going to show up and meet people in the industry, I can't take you seriously.

When I was a teenager, I went to a trade show called ad:tech in San Francisco, and that's where I met my business partner, Harrison. I sat next to him at a sponsored dinner, and we just hit it off. Dinner was kind of lame, so we left and went up to his penthouse at the Mandarin Oriental. He was 14! Don't ask me to explain how any of that was happening, but it was awesome. That was 15 years ago. At this point, we've walked through the fires of hell together, and we've soared higher than either of us ever imagined. It's a relationship I never would have found if I hadn't gone to that trade show.

Maybe it makes me lame, but I've met all my amazing friends and partners at trade shows. It's the only place I know where you can meet good people, talk with them, dine with them, and eventually do business with them. Like, you just can't do that in a bar or on some social media platform.

The first show I ever went to I met this super-famous CEO who happened to be the subject of a popular book I had just read. I walked up to him just to say that I loved his book. We started chatting, and at some point, I mentioned that I was new in the industry and didn't really know anything. He was like, "Cool, come sit with me over here…"

We sat down at a table behind the booth, and he took out a piece of paper and proceeded to teach me everything he knew. I was like a little kid in school learning from a master. I was blown away. Like, what the hell? I just showed up at a show, and here's this famous guy giving me training that most people would pay thousands of dollars for. I couldn't believe it was happening.

The crazy thing is, we went on to do millions of dollars in business with him. I will be forever grateful for the time he spent with me. He was a tough cookie and wouldn't have done that for someone who thought they knew it all. I think maybe that experience is why I attend every show and go out of my way to help people who truly want to learn and grow as performance marketers.

And so that's my biggest tip for you: When you're at a show, don't lie about your numbers. Don't pretend you know more than you do. Help people if you can. But mostly, ask questions and pay attention. Because you never know who might just hand you a million-dollar lesson for free.

I started attending trade shows when I was completely broke and walking around Detroit doing door-to-door sales with a giant hole in my shoe. I knew there was something better in store for me, but I just couldn't figure out what it was.

One day, I went to this giant trade show, and I went to every single booth. I kid you not, every single booth in the exhibit hall—there were probably 200 of them. At each one I said, "Hey, my name's Adam. I'm an affiliate. What do you guys do?"

And then we'd have a conversation. It was exhausting, for sure, but I was motivated. But by the time I was done, I had met people at 200 of the most important companies in the industry. I met a whole bunch of CEOs and executives at these companies. I had opened the door to all sorts of campaigns and promotional methods and opportunities that I could never have imagined existed.

My mind was absolutely blown. All I could think was, *Oh my God! There's so much opportunity here. How do I even pick which ones to pursue?*

Then I went to a bunch of industry parties that people invited me to, and I met more CEOs and a bunch of celebrities. It was absolutely crazy. I stayed in a crappy hotel, spent a little money on the trade show pass, and yet the doors were completely open to me.

One day I'm destitute, walking around Detroit freezing my ass off trying to sell phone equipment, and the next I'm staring at like 50 million dollars' worth of opportunities. Later, staring at a giant pile of business cards, all I could think was, *How do people not know about this? There are just too many ways to make money. This is unbelievable!*

I know some folks try to do all their business at the hotel bar without paying for a pass. This strategy can work in a limited way, but it's not going to put you in front of all the biggest executives in the industry. I see the CEOs of all the companies you want to talk to sitting at their booths and interacting with people. The people controlling the biggest budgets and the most successful campaigns are not sitting at the bar. They're on the show floor meeting and chatting with the folks who *truly* have the desire to succeed.

Have I convinced you to hit the trade shows yet? Good!

Then allow me to give you a few tips so you can have an amazing experience.

1. Look presentable. I can't believe I'm going to share this in a book, but if my embarrassment will save you from the same fate—it's worth it.

Once I started making some money as an affiliate, I went to a trade show covered in diamonds—diamond watch, diamond bracelet, diamond ring. (Definitely not the kind of thing I should have been spending my money on.) But that wasn't the worst of it…

I wore a bathrobe and slippers. Because I thought *I'm an affiliate, I'll do whatever I want. I don't give a crap what anyone thinks!* I looked like an idiot.

Now, the other affiliates thought it was hilarious that I had the balls to do that. And I thought I was *so cool* at the time. But in retrospect, I just cringe. I humiliated myself in front of the very people I wanted to impress. What the hell was I thinking? Today, when I see other affiliates engaging in this same behavior, it really turns me off, and I want nothing to do with them.

Everyone does things in their youth that they regret later on. But when I think about what I might have learned or who I might have met if I had taken myself seriously, it kind of makes me sick. As much fun as I had buying gold chains and fancy cars, I am 100 times more successful since I decided I wanted to develop a real lasting business.

Wear a freakin' suit. Or at least nice jeans and a button-down shirt. If you're a woman, pick a nice outfit that makes you feel professional and confident. Personally, I usually wear nice jeans and sneakers to all the shows now. Our company policy is everyone at the booth wears nice, branded business casual or more formal business attire. Most people choose business casual, shocking I know.

I highly recommend you just skip the embarrassing part and show up looking and behaving like a pro.

2. Bring business cards and a way to take notes. This might sound old-school, but you really need to have professionally designed business cards available to give to people. And you need a way to take notes so you can follow up with people you talk to. That might be voice notes on your phone or a legal pad to scribble notes on. Or maybe you have a business partner who stands by and does the writing.

It doesn't matter how you do it, but you want to follow up with everyone promptly. That's really difficult when you're meeting hundreds of people over the course of a day or two. Most of the time that follow-up falls through the cracks. So few people actually do it, and so many opportunities are just forgotten.

It helps to have a system, and you can feel free to make up your own. Some people put the "must follow up with" cards into one pocket and the ones that don't matter as much in another pocket. Basically, they are sorting out the best opportunities on the spot. Regardless of the opportunities, at a minimum you want to connect with everyone on LinkedIn and follow their social media channels. That way you're keeping the doors to communication open.

You can also take selfies with people to remember their faces and then immediately program them into your contacts on your phone. Of course, they will want a copy of the pic, too, so say "What's your number? I'll send this to you." Now you have their cell number. Texting is a much more personal way to follow up with folks after the show. Don't abuse the privilege, though. Nobody wants to receive a hundred texts from some random person they just met. Just send a quick text, and then take the conversation to email or direct message on social platforms. Be a pro.

3. Have a great time. Be sociable. Be nice to people. You're there to have a good time, and so is everyone else. Go to the parties even if you hate parties. You never know who you'll meet or what kind of amazing doors might open for you. Remember the ice-breaking tips I shared earlier, and just ask people about themselves. *What do you do? What are you working on right now? What's challenging you at the moment? Where do you see the industry going in the next year?* Questions get people talking, and then carrying on the conversation gets easier.

Now, there are a lot of weird personalities at trade shows. Some people are under all kinds of stress and pressure you know nothing about. So if someone tells you no or slams a door in your face, don't sweat it. Some folks are jerks, some are just having a bad day. Some are introverts who want nothing more than to run back to their room and get away from all the people. Just be aware of that, and don't take any negative interactions personally at these events.

4. Don't name drop. If you just met Bill in accounting at the Super Great Gift Card company, don't drop his name unless you say you just met him. If I had gone shooting my mouth off about the famous CEO who spent an hour teaching me privately, it would have gotten back to him, and it could have ruined that relationship and all the opportunity that came from it.

The one exception to this tip is if you see an opportunity for two people to connect. If you're talking to someone who really needs a graphic designer, and you just met a really cool designer an hour ago, by all means connect them. That's giving value to both people and guarantees they'll remember you.

5. Don't lie. I shouldn't have to say this, but I see so many affiliates doing exactly this at trade shows. They lie about their numbers. They lie about who they know and how long they've been in the business. It's so obvious, yet the poor person doesn't know how bad they look. On the flip side, don't undersell yourself either.

I remember sitting in a hotel suite in Bangkok when an affiliate walked into the party we were having. I could tell he was a little freaked out because the room was, like, half the entire top floor of the building. That night, for some reason, I was particularly sick of the affiliates bullshitting their numbers. So, when this guy started talking about how he was doing $50,000 a day, I stopped him and handed him a laptop and said, "Show me or get the hell out."

He looked terrified, but he logged into his tracking system and sure enough, $50,000 a day. It took balls to stand behind his numbers, and now he is a great friend, and we've created a bond that will last a lifetime. That relationship is all based on trust. If his stats hadn't matched his mouth, I would have kicked him out.

Maybe that seems a little harsh, but this shit gets old, and I try to keep my circle small. If someone makes it into that circle, they get tested. If they pass, money can rain from the sky. And you know what? If they tell me they make $20 a day, that's great! Ambition and trust are how mountains get moved. Just don't lie to impress other people. It will backfire.

When I told everybody the truth—that I was a beginner—they bent over backwards to teach me and help me. That's the place you want to be. But if you're projecting an image that you're bigger than you are, people know you're full of it. Just be yourself. It's the best way to go.

CONCLUSION

Congratulations! You made it to the end of the book. That's no small feat. And since you've proven you're persistent, I have a challenge for you.

Your mission, should you choose to accept it, is to take back control of your business and your life. Become a Pay Per Caller and build a healthy, long-term, sustainable business that will take care of you and those you love forever.

I've crammed as much as I possibly could into this book. You have everything you need to get started. It's like you have me looking over your shoulder and directing you what to do next.

But if you'd like my team to *actually* be there to give you feedback on your ads, confirm your routing plans, and answer your questions so you can bypass all the wrong turns and dead ends, then you've got to join the Pay Per Callers forum. It's a vibrant community full of highly successful pay per callers along with folks who are just getting started. We're all in there together building out real businesses. You'll find nearly 50 hours of advanced training from me as well as robust message boards on all sorts of topics. Because you've read this book, you have complimentary access for one month. I want to give you every advantage I possibly can. Scan the code on the resources page to learn how to join.

Once you're inside, please make sure to introduce yourself. My goal is to get to know you and help you out as best I can. But I can't do that if you're just lurking, OK?

Remember that huge gap between my teeth? The one that caused like 15 years of trauma and pain throughout my childhood? Yeah, so once I thought I had officially "made it" in pay per call, I was living in Houston, right down the street from the NASA's Johnson Space Center. It was so cool! You could see the space center right outside my office window. For a geeky kid who went to space camp, it was like a dream.

I remember taking my employees out to a local dive bar one night called Bikini Beach. I wasn't drinking, but we were all having a good time. Then this guy in a nasty stained T-shirt walks up to me and says, "Hey, bro. I saw you pull up in that sick Mercedes. And you got all these people with you. You must be the boss, huh? You're just killin' it at life. But I gotta ask you a question. What the hell is wrong with your teeth? Are you a hockey player or something?"

Oof!

Right in the gut.

I thought I had put all that behind me. I thought who I was would outshine the petty surface judgments. I thought I no longer cared what people thought of me. But sitting there, I was that nerdy kid on the swim team again—the kid everybody called "Gaps."

I woke up the next day and called President Bush's own dentist and drove out to meet him that same week. I didn't know what I was getting into, but that guy was some kind of artist. I told him my story and the whole hockey player comment, and I'll never forget the relief I felt when he said he would help me. It wouldn't be easy, but he would do it.

That treatment was the most expensive thing I'd ever bought besides my car at that point. The procedure took him twice as long as any other veneers he'd done, and it cost me $18,000. But to this day, dentists compliment me on my beautiful smile. When they see the X-rays, they can't believe it. I actually cried once the swelling went down and I finally closed the door on that painful part of my life.

Why am I telling you this? Because that hockey player comment was the moment I decided to take control of my life. That was the moment I decided *no one* was ever going to make fun of my teeth ever again! I didn't care what it might cost, I wanted to feel as confident and successful on the inside as I was on the outside.

I took control of my circumstances and changed my life.

And that's what I invite you to do right now.

Remember what I said at the beginning of this book? There's a revolution coming—a pay per call revolution—and the entire way people buy and sell products and services is about to be turned on its head. Will you be part of the leading edge of that wave? I've given you everything you need to begin.

You're a marketer. You're brilliant at what you do. And even if you're just learning the ropes, you have the dedication it takes to be a massive success.

You deserve better than the uncertainty and instability the traditional affiliate world has to offer.

You deserve to build with your own blocks and keep what you build—not have to start over after some bully kicks your castle down.

You deserve to be rewarded for your persistence and determination.

All of that is possible when you make the decision to take control and change your circumstances. You may not know what you're getting into, but I'm here to help. I won't promise it will be an easy journey, but it will be a hell of a lot more fulfilling than making other people wealthy while you take the scraps.

I know you can do this.

You're ready.

You're set.

Let's go!

PAY PER CALLER MANIFESTO

We believe creative marketing is the most important skill on the planet, and we strive to expand and refine our skill set. No matter how good we get, we know there is more to learn.

We control our own destinies. The best companies in the world seek us out and compete against each other for the traffic we provide.

We present ourselves as professionals, because we are.

We don't build other people's businesses at the expense of our own.

Instead, we build our empires by helping others build theirs.

We work smarter and earn more than any other performance marketer because we have no competition but ourselves.

We believe generating inbound phone calls beats outbound spam every time.

ABOUT RINGBA

Ringba is a call-tracking, attribution-routing, and management platform that was specifically designed for pay per callers regardless of where they sit in the value chain. All of our engineers are ad tech industry veterans who've built some of the most advanced technology in this space. Our goal is to completely change how people buy and sell their calls, and we are absolutely doing that.

We're the fastest-growing platform in the performance marketing space, and we're revolutionizing the way business gets done. Ringba won the G2 customer service award 16 quarters in a row, SaaS Platform of the Year, and the OfferVault Tracking Platform award two years running.

Ringba is set up to be as simple as possible for our clients. We're not the cheapest option for call tracking because we're in the business of helping serious pay per callers make serious money. That means we spend more than anyone else on infrastructure and feature development. Because we know if we can help you build your business and provide you with amazing technology and tools to do it, we'll all be successful.

And so if there's anything we can ever do for you as a performance marketer, please do not hesitate to hit us up. We're very happy to help you in any way we can.

ENDNOTES

Chapter Two

1. Desmarais, Mike. "Call Center Attrition Rate—Is It Now the Most Important KPI?" SQM Group, March 30, 2023. https://www.sqmgroup.com/resources/library/blog/call-center-attrition-rate.

Chapter Nine

1. Statista Research Department. "Digital Ad Spend Worldwide 2026." Statista, March 1, 2023. https://www.statista.com/statistics/237974/online-advertising-spending-worldwide/.

2. Statista Research Department. "Ad-Selling Companies U.S. Digital Ad Revenue Shares 2020–2025." Statista, May 30, 2023. https://www.statista.com/statistics/242549/digital-ad-market-share-of-major-ad-selling-companies-in-the-us-by-revenue/.

3. Bianchi, Tiago. "Global Mobile Traffic 2022." Statista, April 27, 2023. https://www.statista.com/statistics/277125/share-of-website-traffic-coming-from-mobile-devices/.

4. Kolmar, Chris. "26 Surprising BYOD Statistics [2023]: BYOD Trends in the Workplace." Zippia, October 17, 2022. https://www.zippia.com/advice/byod-statistics/.

Chapter Ten

1. "Newspapers Fact Sheet." Pew Research Center's Journalism Project, June 29, 2021. https://www.pewresearch.org/journalism/fact-sheet/newspapers/.

2. "Print Advertising Services." MNI. Accessed July 12, 2023. https://www.mni.com/offline-print-advertising/.

3. Bobnak, Paul. "Average Direct Mail Response Rate: How Effective Is Direct Mail?" Mailing, June 27, 2023. https://www.mailing.com/blog/average-direct-mail-response-rate/.

Chapter Eleven

1. "Annual & Quarterly Revenue." OAAA, May 18, 2023. https://oaaa.org/resources/annual-quarterly-revenue/.
2. "96% of Marketers Achieved ROI Goals with Out-of-Home Marketing Campaigns, Says New Research." *Business Wire*, September 14, 2022. OneScreen.ai. https://www.businesswire.com/news/home/20220914005125/en/96-of-Marketers-Achieved-ROI-Goals-with-Out-of-Home-Marketing-Campaigns-Says-New-Research.
3. Flynn, Jack. "15+ Average Commute Time Statistics [2023]: How Long Is the Average American Commute?" Zippia, June 28, 2023. https://www.zippia.com/advice/average-commute-time-statistics/.
4. *Source: Outdoor Advertising Association of America*

Chapter Twelve

1. Harvey, Steve. "Tune into the USA: The Top US Radio Stations for 2023." Radio Fidelity, January 29, 2023. https://radiofidelity.com/top-us-radio-stations/.
2. https://musicalpursuits.com/radio/
3. Ibid.
4. Ibid.

Chapter Thirteen

1. "Nielsen Estimates 121 Million TV Homes in the U.S. for the 2020-2021 TV Season." Nielsen, July 21, 2022. https://www.nielsen.com/insights/2020/nielsen-estimates-121-million-tv-homes-in-the-u-s-for-the-2020-2021-tv-season/.

ABOUT THE AUTHOR

Adam Young discovered affiliate marketing in his parent's basement while he was going door to door selling phone service instead of going to school. Within a year of quitting his job, he had generated more than a million dollars in commissions. Over the past 20 years, he has become a master of internet marketing, won numerous industry awards, helped thousands of people grow their performance marketing businesses, and co-founded the software company Ringba, which has helped people make billions of dollars through pay per call. He lives in Miami, and you can visit him online at AdamYoung.com.

RESOURCES

There's so much more I want to share to help you succeed in your pay per call journey! Use the code on this page to access some of the awesome free resources I have for you, some of which include:

- Call Back Fraud Examples
- Ringba Demos
- Insertion Order Templates

My team and I are always coming up with exciting new resources for pay per call, so use this code often to see what's new.

GLOSSARY

Abandoned Call (Abandons) - Calls that are not answered before the caller hangs up.

Abandonment Rate - The percentage of callers who hang up before connecting to an agent.

ACL - Average Call Length - When viewing statistics for a campaign, the average call length is represented by taking the total number of minutes and dividing them by the total number of calls. For instance, if your campaign has 10 calls with 600 minutes of total time on the phone, the ACL would be 600 minutes / 10 calls = 60 minutes.

Affiliates (Publishers) - Third parties that create advertisements, promote offers and drive traffic on behalf of brands.

Agent - Person who answers calls in a call center regardless of their job function.

Agent Productivity - A performance metric for determining the productivity of an agent in a call center using factors like abandoned rate, average handle time, after-call work.

AHT - Average Handle Time - The average amount of time an agent is occupied on an incoming call or the average of all time across numerous agents or an entire call campaign.

Allocate Numbers - Assign numbers or issuing new numbers to a number pool, campaign, offer, or publisher.

Blocked Call - Calls that are automatically terminated due to identifying the caller as unwanted or spam.

Busy Signal - A tone used to indicate that the party on the receiving end of a call is currently in another conversion.

Buyer - The party responsible for buying a call. In Ringba Targets can be assigned buyers (owners) with sub-accounts so a buyer can login and see all of their statistics and payouts for all of their targets. Buyers can be assigned to multiple targets because sometimes they may have different partners, queues, or campaigns all in the same brokerage or call center.

CAC (Customer Acquisition Cost) - The cost to acquire a new customer whether that is through advertising directly or through buying calls.

Call Attribution - By linking calls to specific source and user data it allows you to optimize your campaigns and attribute revenue to specific traffic and call sources. Without detailed information about who is and is not calling, it is not easy to optimize campaigns by performance.

Call Center Efficiency - Call Center Efficiency is the process of optimizing the operations of a call center, including; response time, handle time, abandonment rate, average handle time, agent productivity and caller satisfaction.

Call Duration - Length of the call from the moment the call is connected to our server and the moment the call is terminated. Also referred to as 'call length' or the 'length of a call'.

Call Flows - Ringba's system for visually designing the flow of a caller after they call a phone number.

Call Log - In-depth log of all a call's details, meta-data, call recording and any other data associated with that call.

Call Tracking - The process of associating advertising sources with generated calls.

Call Trading (Brokering Calls) - The business of being an intermediary between a buyer and seller of a phone calls.

Glossary

Call Queue - This is where calls are held when they cannot be immediately answered by agents. Queues typically run on a first-in, first-out basis, but can also be configured into priority and waterfall scenarios.

Caller ID - Caller Identification (CNAM - Caller Name) - The phone number and name of the caller.

Campaign - Settings assigned to a specific group of actions, publishers, or targets to meet the needs of your clients.

Capacity - The amount of call traffic, or concurrency, a target can handle.

Capacity Management - The management of human resources to match the number of predicted inbound calls throughout the day.

Caps - Hourly, daily, weekly and/or monthly restrictions on the number of calls a buyer can buy. For instance, a call buyer may only buy up to 25 calls per hour from a seller.

Click to Call - A type of user interaction for connecting with a business or individual.

Completed Call - A call that has hungup and no one is on the line.

Concurrency (CC, cc) - The number of simultaneous calls that a contact center, target, or campaign can receive, or is receiving, at any given time. For instance, if a call center has 25 agents, their maximum concurrency is 25.

Connect - When a caller and an receiving party are connected, whether it be with an agent, or a phone system that answers the call.

Connected - The moment when a call is answered by an agent or automated system.

Conversion – A call that results in some type of financial or payout event. Sometimes these are referred to as 'duration based conversions,' where the caller stays on the line for a specific length of time before a payout or revenue event is triggered.

Conversion Rate - The percentage of calls that converted. This is calculated by taking the number of conversions and dividing them by the number of calls. For instance, 10 conversions divided by 100 calls would result in a 10% conversion rate.

CPA/CPL - The Cost Per Action or Cost Per Lead (Call) is the amount of money that it costs to generate a conversion event. That conversion event could be a click, lead, call, sale, or some other kind of activity that a user takes.

Deallocate (Deallocate Number) - Remove a number from your Ringba account. Once a number is deallocated, it is released back to the phone carrier and cannot be repurchased or added back to your account.

DID - Direct Inward Dial - A virtual phone number used to route inbound calls. All phone numbers issued inside of Ringba are DIDs.

Disposition (Call Disposition) - The outcome or result of a phone call. Example: Sale, not interested, follow-up.

Dropped Call - A phone call that is terminated or disconnected abruptly before the speakers finished their conversion due to some kind of technical difficulty, not the users hanging up the phone.

Duplicate Call - A repeat call made by the same caller within a specific period of time.

Dynamic Number - A unique phone number that is passed back to a Ring Tree with a bid.

EPC - Earnings Per Call (RPC) - Estimated revenue per call generated for a campaign. Typically calculated by taking the revenue and dividing it by the number of calls it took to generate.

Failed to Connect - When a target's system is unable to answer or connect a call it is marked as a failure.

Geo Routing - The routing of a calls based on the caller's actual geographic location.

Granularity (Reporting Granularity) - Granularity refers to the specificity of reporting information that is available inside tracking platforms.

Examples being the caller's: Device, operating system, geographic location, carrier, landing page, and more.

Handle Time - The length of time a caller spends talking to an agent.

Hold Time - The length of time a caller spends waiting on hold to speak to an agent. Hold time is calculated from the moment the call is connected to the moment when an agent answers the call.

Hours of Operation - The hours that a contact center can accept calls.

Inbound Call - A call that is originated by an external source, or caller.

Inbound Call ID - The unique identifier that is assigned to every inbound call in Ringba. Not to be confused with a 'Caller ID.'

Inbound Number - A phone number designated to receive calls.

Insights - Ringba's analytics system that allows users to segment and review their reporting data to find new opportunities.

Instant Caller Profiles (ICP) - Ringba's real-time data service that appends all kinds of data to a caller ID that you can then use to make bidding and routing decisions, such as caller name, address, financial data, mortgage and property data, and more.

IVR - Interactive Voice Response - A system responsible for interacting with callers by requesting they press numbers on their dial-pad or speak into the phone.

Leg - A segment of the entire network-to-network path that a call is routed through. For instance when someone references an "outbound leg", they're speaking about the route from their system or platform to the end user. An "inbound leg" refers to the opposite, the route from the caller to their system or platform.

Load Balancing - The process of optimizing how calls are routed based on capacity and/or performance.

Local Number - A type of phone number that is associated with a specific geographic location.

Misdials - Connected calls where the caller dialed an incorrect number.

MMS - Multimedia Message System - The technical term for a text message that includes multimedia files such as pictures, audio or video.

Mobile Number - The phone number of a mobile phone or device that connects to the network wirelessly.

Mobile Number Portability - The ability to re-assign a mobile phone number to a different carrier.

No Connects - Calls that were initiated but dropped or unanswered by their destination targets.

No Target Answered - When a call is not answered by any of a campaign's targets Ringba throws this error event.

No Target Found - When there are no targets found based on the routing criteria for a call Ringba throws this error event.

Number Format - The format used to display a phone number generated by a Ringba JavaScript tag. You can specify the format inside your account and Ringba will automatically format your dynamic numbers visually the way you prefer. For instance (nnn) nnn-nnnn.

Number Lock - When a number from a number pool is assigned to a user, Ringba locks it for a specific period of time allowing the user to call.

Number Pool - A group of phone numbers used for caller analytics.

Number Pool Misses - This happens when all numbers inside of your number pool are in use by other sessions. When you start to see this warning you need to increase the size of your pool.

Optimizing Call Flow - The process of improving the return on investment of your efforts by changing settings and rules inside of your campaign configuration to better deliver calls to buyers. Also called Call Flow Optimization.

Outbound Call (Outbound) - A call originated by an agent to reach a customer outside of their contact center or location.

Overflow - Excess call traffic that is greater than a call center's capacity.

Paid Call - A call that meets conversion criteria and pays commission.

Payout - The money paid to your publisher for every conversion that meets a campaign's criteria.

Pay Per Call (PPCall) - An advertising, billing, and performance marketing model for connecting businesses with inbound customer calls.

Pay Per Call Network - A platform that manages Pay Per Call campaigns on behalf of advertisers and generates calls through the network of publishers.

Pay Per Callers - People involved in the Pay Per Call industry in some way call themselves "Pay Per Callers". It is also the name of the industry's only forum you can find at PayPerCallers.com.

Per Minute Usage - The associated cost for a single minute of voice traffic through a platform or provider.

Postback/Postback Pixel - An easy way to pass data to other software platforms by pinging them during and after phone calls.

Predictive Dialer (Dialer) - An outbound calling system that automatically dials from a list of telephone numbers. Also known as; robodialer, autodialer

Predictive Routing - Ringba's automated routing engine that predicts the highest value destination of a phone call and automatically routes it.

Priority - A scoring variable used by Ringba to determine which target to send a call.

QA - Quality Assurance - The process of determining whether a call or agent's performance meets specified requirements.

Raw Call - An inbound call, whether answered or not. This term is typically used to describe a billing method where the buyer is responsible for paying for every call, whether they answer it or not. Example: "$10 per raw call"

Real-Time Bidding (RTB) - Ringba's bidding system that allows buyers to bid on phone calls based on actual consumer demographic information.

Recording (2-Channel Recording, Call Recording) - The act of recording and storing an audio file of a phone call. Two channel recordings record each party on a call inside of a separate audio channel in a single file and you can hear them separately on the right and left speakers.

Repeat Call (Duplicate Call) - A call from someone who has previously called in.

Reporting - Analytics and tools for reviewing call traffic, call logs, caller details and recordings.

Ring Trees® - Ringba's real-time integration system that allows call sellers to instantly integrate with call buyers that bid on phone calls.

Ringba JavaScript Tag (Ringba Tracking Tag, Ringba Tag) - A snippet of code used by Ringba to track information about users and to display phone numbers on web pages.

Ringless Voicemail - A method of leaving pre-recorded audio messages in someone's voicemail without their phone ringing. Also known as a 'voicemail drop' or 'Ringless Voicemail Drop' (RVD or RVM).

Robocall - An automated call done by a computer that plays back a pre-recorded message to the user.

Routing Plan - A collection of targets for managing calls flows and routing call traffic inside of your call tracking platform.

RPC - Revenue Per Call - The gross revenue generated per inbound call to a call center. This is calculated by taking the gross revenue and dividing it by the number of calls it took to generate the revenue.

Self-Reporting - When a party provides reporting information about calls and conversions without full transparency for all parties involved.

Short Duration Calls - Extremely short phone calls that last for 5 seconds or less.

SIP - Session Initiated Protocol - SIP is a VOIP Protocol used by most cloud based phone service providers. Using SIP allows you to take a VOIP call and connect it directly without passing it through the offline telecom system preserving higher voice quality and reducing connection times.

SIP Endpoint - A SIP Endpoint is the connection address used to route a call using VOIP, without connecting to the telephone network. Also known as a "SIP Address".

SIP Header - The header of a packet sent using the SIP protocol. This header can contain information about the call, unlike a traditional call, that allows you to attribute it to specific data similar to a traditional online marketing tracking link.

SMS - Short Message Service - The technical term for a text message (txt message). These messages have specific length restrictions based on country and carrier.

Speech-to-Text - Assistive technology that uses computers to translate spoken language into text. Also known as "Transcription".

Sub-Account - Employee or team member account that is controlled by your main account to allow permission-based access to different areas of Ringba.

SubID - Sub Identification - An identification variable that tracks specific information about a user. Typically used with keywords, traffic sources or some other kind of data to attribute the user's actions to a specific source.

Tag Routing - The ability to dynamically route calls using tags or data points inside of Ringba.

Target - The receiving party of a call. Usually this is a salesperson, call center, marketplace, or broker.

Target Number - The phone number or SIP address of a target that will be used to route calls to them.

Target Priority - The priority of a target determines what order a call will be routed if there are multiple targets that are open and have available concurrency.

Target Timeout - The amount of time that a target has to answer a call before it is routed to another target.

Target Weight - A sub-setting of the target priority, weight allows you to add a target to a campaign and adjust its relative priority within a group without re-organizing numerous other targets. Weight is calculated after priority.

Targets - A destination or phone number that inbound calls can be routed to.

TCL - Total Call Length - The total call length for a specific campaign is a total of the time spent on the phone for every call over a specific period of time. For instance if you had 10 calls that were each 30 minutes, you would have a total call length of 5 hours. To calculate this we take 30 minutes * 10 calls for 300 minutes, divided by 60 for a conversion to hours, is 5 hours of total call length.

Text-to-Speech - Assistive technology that uses computers to read text aloud.

TFN - Toll Free Number (Telephone Number) - Type of phone number where the receiving party pays for the telecom costs. This is mostly irrelevant now in the United States and Canada with the advent of unlimited calling plans, however we find Toll Free Numbers have a much higher conversion rate as they are associated with trust.

Time to Call - The time between when a user is shown a phone number and when they call. This is calculated by subtracting the time when the user was first shown the phone number from the time when they actually called.

Time to Connect - The time between when a call is accepted into Ringba and the moment it is connected with a target.

Transcription - The process of converting speech to text

VOIP - Voice Over Internet Protocol - Phone and voice services that happen over the internet without requiring a user to use traditional 'copper wire' phone lines. Many international calls are routed over VOIP to reduce the transport cost of the call. Services like Google Voice and Skype use VOIP to connect to the telephone network and allow their users to make calls.

Warm Transfer - A type of transfer where an agent speaks to the person they transferring their call to prior to handing off the call.

Webhook - A way to integrate that allows third parties to pass revenue and other data back into Ringba so it can be viewed in reporting.

Welcome Message - The recording or text-to-speech message a caller hears when their call is initially answered.

Whisper Message - A message played back to the receiving party of a call before it is connected to give them information about the call or caller.

INDEX

NUMBERS

1-800-DENTIST, 189
65+ demographic, ad for, 105

A

A4D affiliate network, 63
abuse, experience of, 9
ad approval requirement, 84–85
ad campaigns, building, 5
ad copy, 103–105, 110–112, 142
ad samples. *See also* print advertising
    65+ demographic, 105
    people in their 20s, 106
ads. *See also* campaigns; paid ad space
    creating and testing, 106
    creating for print advertising, 142–143
    getting opinions of, 75
ad:tech trade show, 247
advertising, purpose of, 101–102
advertising costs, payment of, 44–45
advertising strategies, 20
advertising team, 41–42
AdWords, 76–77
affiliate model
    appeal of, 5
    realities of, 6
    traditional example, 7
affiliates
    bringing on, 38
    building relationships with, 244
    communicating caps to, 204
    direct mail, 146
    failure of, 241–245
    leveraging to scale, 202
    success of, 79
    vetting, 204
    weakness of, 47
agent cherry picking, 92–93. *See also* sales agent performance
agreements, understanding, 86–87. *See also* contracts
AI writing tools, using, 127
Akatiff, Jason, 63
Amazon digital radio, 162
Amazon paid ad space, 114
Amazon Prime, 185
American phone number, acquiring, 69
AM/FM radio, 162–163
annoyance, root of, 98
anxiety
    association with debt, 107
    fear and panic, 104–105
    seeking relief from, 108
AOL chat rooms, 2
Apple digital radio, 162
assets and brands, creating, 81–82. *See also* professional brand
AT&T radio advertising, 163
Audacy audio advertising research, 163

B

Barnes and Noble, doing market research in, 138–139
bidding for phone calls, 236–239
billboard advertising
    ads, 156–157

location, 155–156
media vendors, 156
phone numbers, 156
rates, 158–160
results, 157
blind price negotiation, 210
blog posts and SEO, 127
brands and assets, creating, 81–82. *See also* professional brand
broadcast TV, 176
broker flow, 30–31
B-roll footage, using for commercials, 183–184
budget versus payout, 79. *See also* testing budget
bullying, experience of, 9
business checking account, setting up, 70
buyer network
  call pricing, 196
  contacting prospects, 198–200
  creating, 190–193
  deciding on markets, 195–196
  leveraging affiliates to scale, 202
  listing potential buyers, 196–197
  self-serve, 193–195
  starting, 202–204
  vetting affiliates, 204
buyers
  dialing and selling to, 222–224
  pay per call campaigns, 17
  rerouting automatically, 219–221
  working with, 200–202
buying behavior, Campaign Avatar Profile Creator, 137
BYOD (bring your own device) policy, 115

C

cable TV, 176
call centers
  environments of, 15–16
  going after, 37
call flows
  direct buyer, 26
  IVR input qualification, 27–28
  multiple brokers, 30–31
  network, 29–30

  optimizing, 45–47
  prioritizing and weighting, 218–219
  understanding, 25
  warm transfer, 28–30
call quality, negotiating with, 214–216
call recordings, listening to, 17
call routing
  basing on concurrency, 233–235
  basing on hours of operation, 232–233
  duplicating, 224–225
  predictive, 239–240
  real-time bidding, 236–239
call to action, print advertising, 142
call tracking
  importance of, 31–33
  setting up, 117
callback fraud and duration billing, 87–90
call-only advertisements, 123–124
calls. *See also* pay per call; "qualifying calls;" unanswered calls
  answering, 31
  buying intent of, 14–15
  getting credit for, 21
  getting paid for, 86
  listening to, 79–80, 83
  real-time bidding for, 236–239
  tracking phone number, 20–22
Campaign Avatar Profile Creator, 136–138
campaigns. *See also* ads
  beginning, 74–75
  brands and assets, 81–82
  choosing traffic sources, 78
  establishing testing budget, 75–78
  load balance calls, 80–81
  promoting to publishers, 18
  relationships, 80–81
  starting testing, 79–80
  testing, 118–120
cc (call flow concurrency), routing based on, 233–235
challenges and pain points, Campaign Avatar Profile Creator, 136
classified ads, 20, 142–143
Clear Channel, out-of-home advertising, 152
click-through rates, monitoring, 119
click-to-call advantage, 121–122

CMYK versus RGB colors, print advertising, 143, 157
.com domain, buying, 81, 127
Comcast radio advertising, 163
commercials
 audio, 183
 B-roll footage, 183–184
 editing, 184–185
 iPhone and gimbal, 183
 lighting kit, 183
 message delivery, 184
 scripts, 183
 television advertising, 180–184
commissions
 generating, 20–22
 maximizing amounts of, 24
communication
 importance of, 41
 reading aloud, 53
communication channels, Campaign Avatar Profile Creator, 137
company organization, members of, 42
competition, awareness of, 116
compliance, 40, 84
concurrency, routing based on, 233–235
consumers, questions asked by, 16–17
contact information, making available, 53–54
contracts, brand-building basics, 64–67.
 See also agreements
conversion duration, 22–23
conversion rates, calculating, 23–25
conversion strategy, Campaign Avatar Profile Creator, 138
copywriting, 103–105, 110–112, 142
corporation or LLC, setting up, 69–71
cost of acquisition, calculating, 15
COVID, 189
creatives, split testing, 118–120
curiosity, using as psychological tool, 95–96
customer journey, Campaign Avatar Profile Creator, 137
customers, calculating cost of acquisition, 15

D

debt consolidation example, 107–108
demographic routing, 228–230
demographics, Campaign Avatar Profile Creator, 136
Dentist scenario, 24–25
design logos, investing in, 127
despair, association with debt, 107
digital designers, 157
digital radio, 161–162
digital traffic sources. *See* online advertising
direct buyers, 26, 36–37
direct mail, 132, 145–147
DMA (designated market area), television advertising, 186
duplicate call routing, 224–225
duration billing and callback fraud, 87–90

E

Edison Research, AM/FM radio, 162–163
editing software, using for commercials, 184
EIN (Employer Identification Number), getting, 70
email
 online advertising, 125–126
 professionalism of, 69
 spam, 2–3
emotion
 evoking through images, 112
 triggering, 96
emotion words, negative and positive, 109
emotional cause and effect, 99–102
emotional combinations, 102–103
engagement triggers, Campaign Avatar Profile Creator, 138
enterprise value, creating, 37–38
experience, positive versus negative, 96

F

Facebook ads, 76, 114
Facebook affiliates, 10
Facebook business page, setting up, 69
Facebook groups, 56–57
failures, confronting, 8
fake calls, 90
favoritism, acknowledging, 43
fear
 association with debt, 107

panic and anxiety, 104–105
  role of, 101
feelings, storage as data, 102
Final cut Pro editing software, 184
Fiverr, 119, 143
fixed-call cost, 15
Fox News, target audience, 179
fraud, checking for, 86
function, role in emotional cause and effect, 101
funds, mismanagement of, 43
Futuredontics, 189

G

Geocities, 2
geographic routing, 225–228
geographical load balancing, 46–47
goals for campaign, Campaign Avatar Profile Creator, 137
G.O.A.T. Club, 189
Google, call-only advertisements, 123–124
Google AdWords, 76–77
Google Business Profile, 128
Google keywords, expansion of, 49–50
Google Maps, using for potential buyers, 196
Google's market share, 114
guilt, explanation of, 102

H

Home Depot radio advertising, 163
hoplessness, association with debt, 107
hosting package, early example of, 2
href, setting to URL, 121
HTML link, example of, 121
Hulu, 185–186
human experience, positive versus negative, 96

I

idle time, avoiding, 44
iHeard Radio, 162
iHeartMedia, out-of-home advertising, 153
industry rules, playing by, 40. *See also* verticals

iPhone and gimbal, using for TV commercials, 183
IVR (interactive voice response), 27–28, 230–232

J

JCDecaux, out-of-home advertising, 153

K

keywords
  cost-per-click, 78
  rethinking, 79–80

L

Lamar, out-of-home advertising, 152
landing pages, 74–75, 118
lawyers, using services of, 64–67
lazy marketers, 116
LegalZoom, 70
libraries, doing market research in, 139
lighting kit, using for commercials, 183
link building, SEO, 128
LinkedIn groups, 57
LinkedIn profile, creating, 68
LLC or corporation, setting up, 69–71
load balance calls and relationships, expanding, 80–81
local search, online advertising, 128–129

M

marketer
  being, 4–5
  lazy type of, 116
  as memory miner, 96–97
marketing alternatives, 10
Matchstick Legal, 70
media, buying, 76
media and information consumption, Campaign Avatar Profile Creator, 136
media and news channels, 103
media kits, researching for publications, 139–140
memory miners, marketers as, 96–97, 105
message, Campaign Avatar Profile Creator, 137

Meta. *See* Facebook ads
mobile accounts, web traffic, 114
money, budgeting, 77
motivation and desires, Campaign Avatar Profile Creator, 136
multiple brokers flow, 30–31
music types vs. demographics, 171
Musical Pursuits.com, 162
mystery shoppers, 90

N

native, online advertising, 126
negative emotion words, 109–112
negative events, trauma associated with, 102–103
negotiating financials, 41
negotiating for scale
    call quality, 214–216
    ground rules, 206–209
    network vs. network, 210–214
    overview, 205
Netflix, 185
network flow, 29–30
network relationships
    removing dependence on, 37
    starting with, 36
network vs. network, 210–214
networks. *See also* pay per call networks
    problem with, 21
    starting, 202–204
news channels and media, 103
newspapers, reach of, 132

O

objection handling, Campaign Avatar Profile Creator, 138
offers
    evaluating, 60–61
    picking, 5
OfferVault website, 57–58, 61
offline marketing, 10
online advertising. *See also* traffic
    email, 125–126
    local search, 128–129
    native, 126

paid search, 123–124
and pay per call, 117–121
pros and cons, 115–117
push notifications, 125
SEO, 126–128
social media, 124
spending worldwide, 114
online marketing, 9–10
OOH (out-of-home) advertising
    overview, 149–153
    pros and cons, 153–155
OTT (over-the top) and streaming platforms, 185–186
OUTFRONT Media, out-of-home advertising, 152

P

paid ad space, biggest players in, 114. *See also* ads
paid search, online advertising, 123–124
Pandora digital radio, 161–162
panic, fear, and anxiety, 104–105
pay per call. *See also* calls; "qualifying calls;" unanswered calls
    math, 22–23
    overview, 6–11
    path to success, 36–38
    steps, 17–22
    superiority of, 14–17
    upward spiral, 9
Pay Per Call Masterclass, 141
pay per call networks. *See also* networks
    call flow optimization, 45–47
    function of, 38–42
    getting started, 184–184
    pros and cons, 42
    working with, 53–60
Pay Per Caller Manifesto, 257
Pay Per Callers forum, 75, 93, 141, 174, 183, 204
payout versus budget, 79
PayPerCallers.com, 56
phone calls. *See also* pay per call; "qualifying calls;" unanswered calls
    answering, 31

284   The Pay Per Call Revolution

buying intent of, 14–15
getting credit for, 21
getting paid for, 86
listening to, 79–80, 83
real-time bidding for, 236–239
tracking phone number, 20–22
phone number
　acquiring, 69, 134
　checking, 146
platforms and verticals, 117–118
Plutchik's Wheel of Emotion, 97–98
Pocket Your Dollars, 35
positive emotion words, 109–112
predictive routing, 239–240
price negotiation, 210
print advertising. *See also* ad samples
　call to action, 142
　Campaign Avatar Profile Creator, 136–138
　creating ads, 142–143
　direct mail, 145–147
　finalizing placement, 143
　negotiating price, 140–141
　overview, 131–135
　"rack rates," 140–141
　researching playing field, 138–140
　resolution, 143
　target market, 135
　tips, 143–145
professional brand. *See also* assets and brands
　building, 38, 63–67
　creating, 67–69
　LLC or corporation, 69–71
professional presentation, 53
prospects, contacting, 198–200
psychographics, Campaign Avatar Profile Creator, 136
psychology of advertising, Plutchik's Wheel of Emotion, 97–98
PTSD, problem of, 102–103
publications, researching media kits, 139–140
publisher team, volume consistency, 41
publishers
　committing fraud, 90–91

inbound calls, 19–20
promoting campaigns to, 18
push notifications, online advertising, 125

Q

QA (quality assurance), 40, 89
QR codes
　callback fraud, 87–88
　call links and OTT, 186
QR to call, 122
"qualifying calls," 21–22. *See also* calls; pay per call; unanswered calls
quality, monitoring, 41
quality assurance, 85–87

R

"rack rates," print advertising, 140–141
radio ads
　activities of listeners, 172
　failures, 173–174
　foreign language stations, 172
　formats, 170–172
　music types vs. demographics, 171
　phone number, 170
　PSAs (public service announcements), 169–170
　recording, 169
　volume discounts, 172–173
radio advertising
　ad creation, 167
　basics, 165–168
　geographic targets, 165
　overview, 161–163
　price negotiation, 167
　pros and cons, 163–165
　radio stations and ad reps, 166
　results, 168
　spot length and placement, 166–167
　target audience, 165
　time slots, 168
　tips, 169–174
relationships and load balance calls, expanding, 80–81
research
　keyword cost-per-click, 78
　performing for verticals, 52

RGB versus CMYK colors, print advertising, 143, 157
Ringba call-tracking platform
  description, 259
  landing pages, 74
  optimizing call flow, 45
  PayPerCallers.com, 56
Ringba secrets
  call flow, 218–219
  demographic routing, 228–230
  dialing and selling to buyers, 222–224
  duplicate call routing, 224–225
  geographic routing, 225–228
  IVR selection/qualification routing, 230–232
  overview, 217–218
  predictive routing, 239–240
  real-time bidding for phone calls, 236–239
  rerouting busy buyers, 219–221
  rerouting unanswered calls, 221–222
  routing and concurrency, 233–235
  routing and hours of operation, 232–233
risk tolerance, setting, 77
routing
  basing on concurrency, 233–235
  basing on hours of operation, 232–233
  duplicating, 224–225
  predictive, 239–240
  real-time bidding, 236–239
royalty-free, explained, 184
RPC (revenue per call), calculating, 23–24
rules, playing by, 40, 84

S

sales agent performance, poor quality of, 91–92. *See also* agent cherry picking
sales conversations, listening to, 16–17
Sarandrea, Anthony, 35
satellite radio, 161–162
satellite TV, 176
scripts
  noncompliant warm transfer, 90–91
  using for training programs, 201–202
  writing, 194

SEO, online advertising, 126–128
services, selling, 197
shame, association with debt, 107
Shark Tank, 22
"simuldialing" buyers, 222–224
social media, online advertising, 124
social platform, creating, 81
social profiles, editing, 55, 68–69
SoundCloud, 161
spam, 2–3
Spanish-speaking audiences, 61–62
Spotify, 161–162
star ratings and reviews, recording, 197
statistics, accuracy of, 21
stimulus event, role in emotional cause and effect, 100–101
streaming platforms and OTT, 185–186
subconscious appraisal, role in emotional cause and effect, 101
survivor spirit, 8

T

television advertising
  choosing markets, 181–182
  commercials and tracking numbers, 180–181
  commercials on budget, 182–184
  distribution areas, 179–180
  DMA (designated market area), 186
  editing commercials, 184–185
  flight dates, 181
  negotiating rates, 185
  networks, mediums, and brokers, 180
  OTT and streaming platforms, 185–186
  overview, 175–177
  placement ROI, 181
  placements and flight dates, 180
  pros and cons, 177–179
  rates, 179
  running, 77
  schedules, 179–180
  shows, 179–180
  steps, 179–181
  target audience, 178
  testing, 178
  viewer statistics, 181

YouTube, 186–188
testing
    importance of, 241
    starting, 79–80
testing budget, establishing, 75–78. *See also* budget versus payout
testing campaigns, 118–120
third-party traffic, rule about, 85
T-Mobile radio advertising, 163
tracking, importance of, 32–33
tracking phone number, 20, 33
tracking platforms, 21
trade shows, tips, 58–60, 247–252
traffic. *See also* online advertising
    buying, 77
    sources, 78
    third-party, 85
transparency
    and compliance, 84
    lack of, 43

U

unanswered calls, rerouting, 221–222. *See also* calls; "qualifying calls;" unanswered calls
Upwork, 119, 143
URL, setting href to, 121
USP (unique selling proposition, Campaign Avatar Profile Creator, 137

V

value proposition, Campaign Avatar Profile Creator, 138
verticals. *See also* industry rules
    choosing, 50–52, 243
    evaluating offers, 60–61
    pay per call networks, 53–60
    and platforms, 117–118
    Spanish-speaking audiences, 61–62
video ads, 119, 127
video-streaming sources, 185
volume consistency, publisher team, 41

W

warm transfer flow
    explained, 28
    via network flow, 29–30
    noncompliance, 90–91
website, building for professional brand, 68, 81
World Wide Web, making money on, 2

X

XM radio, 161

Y

YouTube, 186–188

TESTIMONIALS

Pay per call is a golden opportunity that turns traditional affiliate marketing on its head. Higher returns for the advertisers, better experience for the consumer, and more money for the performance marketer—it's a win-win-win scenario, creating real value for everyone involved.

If you're a marketer who's tired of networks and advertisers controlling your cash flow, you owe it to yourself to check out pay per call. You call the shots. You decide how much to charge for your calls. You make the lion's share of the revenue. There's no other business like it!

—Dave Maman
CEO of WeCall

In the vast, exciting world of marketing, there's a goldmine just waiting to be discovered: the pay per call industry. It's an open field, ripe with opportunity for marketers who are ready to leverage their skills and drive consumer phone calls straight to the big-name brands. The amazing part is, most affiliates aren't even aware it exists.

Now is your chance to jump in and build something monumental. We're talking about creating a business that's not just here today, gone tomorrow, but one that stands the test of time, built solidly on your terms.

—Manny Zuccarelli
CEO of Quote Velocity

Pay per call isn't just about driving clicks; it's about helping consumers connect in real time to real businesses that can help them. That's real value! We're elevating the whole marketing game to levels you didn't even think were possible. This is the golden ticket, the opportunity that will set you apart from the pack so you can build the business and life you're dreaming of.

—Yohan Perez
CEO of Rank Media